AsiaWorld

Series Editor: Mark Selden

This series charts the frontiers of Asia in global perspective. Central to its concerns are Asian interactions—political, economic, social, cultural, and historical—that are transnational and global, that cross and redefine borders and networks, including those of nation, region, ethnicity, gender, technology, and demography. It looks to multiple methodologies to chart the dynamics of a region that has been the home to major civilizations and is central to global processes of war, peace, and development in the new millennium.

Titles in the Series

China's Unequal Treaties: Narrating National History, by Dong Wang

The Culture of Fengshui in Korea: An Exploration of East Asian Geomancy, by Hong-Key Yoon

Precious Steppe: Mongolian Nomadic Pastoralists in Pursuit of the Market, by Ole Bruun

Managing God's Higher Learning: U.S.-China Cultural Encounter and Canton Christian College (Lingnan University), 1888–1952, by Dong Wang

Queer Voices from Japan: First Person Narratives from Japan's Sexual Minorities, edited by Mark McLelland, Katsuhiko Suganuma, and James Welker

Yōko Tawada: Voices from Everywhere, edited by Douglas Slaymaker

Modernity and Re-enchantment: Religion in Post-revolutionary Vietnam, edited by Philip Taylor

Water: The Looming Crisis in India, by Binayak Ray

Windows on the Chinese World: Reflections by Five Historians, by Clara Wing-chung Ho

Tommy's Sunset, by Hisako Tsurushima

Lake of Heaven: An Original Translation of the Japanese Novel by Ishimure Michiko, by Bruce Allen

Imperial Subjects as Global Citizens: Nationalism, Internationalism, and Education in Japan, by Mark Lincicome

Japan in the World: Shidehara Kijūrō, Pacifism, and the Abolition of War, Volumes I and II, by Klaus Schlichtmann

Filling the Hole in the Nuclear Future: Art and Popular Culture Respond to the Bomb, edited by Robert Jacobs

Water

The Looming Crisis in India

Binayak Ray

LEXINGTON BOOKS

A division of
ROWMAN & LITTLEFIELD PUBLISHERS, INC.
Lanham • Boulder • New York • Toronto • Plymouth, UK

LEXINGTON BOOKS

A division of Rowman & Littlefield Publishers, Inc.
A wholly owned subsidiary of The Rowman & Littlefield Publishing Group, Inc.
4501 Forbes Boulevard, Suite 200
Lanham, MD 20706

Estover Road
Plymouth PL6 7PY
United Kingdom

Copyright © 2008 by Lexington Books
First paperback edition 2010

All rights reserved. No part of this publication may be reproduced, stored in a retrieval system, or transmitted in any form or by any means, electronic, mechanical, photocopying, recording, or otherwise, without the prior permission of the publisher.

British Library Cataloguing in Publication Information Available

The hardback edition of this book was previously cataloged by the Library of Congress as follows:

Library of Congress Cataloging-in-Publication Data

Ray, Binayak.
   Water : the looming crisis in India / Binayak Ray.
     p. cm. — (AsiaWorld)
   Includes bibliographical references and index.
   1. Water-supply—Government policy—India. 2. Water resources development—India. 3. Water resources development—Government policy—India. 4. Water-supply—India. I. Title.
HD1698.I4R39 2008
333.9100964—dc22   2007045423

ISBN: 978-0-7391-2601-1 (cloth : alk. paper)
ISBN: 978-0-7391-2602-8 (pbk. : alk. paper)
ISBN: 978-0-7391-3027-8 (electronic)

Printed in the United States of America

∞™ The paper used in this publication meets the minimum requirements of American National Standard for Information Sciences—Permanence of Paper for Printed Library Materials, ANSI/NISO Z39.48-1992.

To the memories of my parents, the late Dhirendreswar and Smriti Kana Roy, and the late Baneswar Roy, Pradyut Kumar Pal, and Sunil Sen Sharma.

Contents

List of Tables — ix
Foreword by Vandana Shiva — xi
Preface — xiii
Introduction — xvii
Abbreviations — xxv

Part I: The Broader Context

Chapter 1 The Water Environment — 3
Chapter 2 Water Policy: An Anatomy — 23
Chapter 3 International Rivers: Global Conventions, Regulations, and India — 49
Chapter 4 Future Demand for Water and Available Options — 63
Chapter 5 The River-Linking Project — 89

Part II: Issues for India and the Region

Chapter 6 Environmental Perspectives — 105
Chapter 7 Economic and Financial Perspectives — 115

| Chapter 8 | Political and Governmental Perspectives | 127 |
| Chapter 9 | Regional Perspectives: Bangladesh, Nepal, Pakistan, and China | 141 |

Part III: The Way Out

Chapter 10	Reflections	149
	Appendix A: Insurgency and Water: Three Cases	165
	Appendix B: A Template for an EIS	171
	Appendix C: Sociopolitical environment in Islamic Republic of Bangladesh, People's Republic of China, Royal Kingdom of Nepal, and Islamic Republic of Pakistan	175
	References	189
	Further Reading	221
	Index	225
	About the Author	231

List of Tables

1.1	Renewable freshwater by region.	5
1.2	Current annual freshwater withdrawals by four South Asian countries by sector.	11
1.3	Annual freshwater withdrawal by sector and income group.	12
1.4	Percentage of major river basins in sub-continental countries and China.	14
1.5	Freshwater flows and per capita water available in mainland South Asia in 2000.	15
1.6	Actual and projected population in the mainland sub-continental countries: 1980–2015.	17
2.1	Violation of pollution standards in some main Indian rivers.	27
2.2	Estimated freshwater savings in agricultural sector by investing in water-saving technology.	33
4.1	Estimated annual per capita freshwater withdrawal in South Asian countries and China by purpose in 2000.	65
4.2	Per capita domestic water use in 2000 in mainland South Asian countries and China.	65
4.3	India's estimated freshwater needs to 2050 by sector under alternative growth assumptions.	66
4.4	India's estimated freshwater needs to 2020 by sector under different scenarios.	66

4.5	Siltation in selected Indian river-dams: projected and actual (in acre feet per annum) and its impact on dams' life expectancy.	73
4.6	Extent of sloping land in India by soil type and class (in million hectares).	73
4.7	Rural population growth and density per km^2 of arable land in South Asian countries.	80
4.8	Changes in the percentage sof arable land and irrigated land in South Asian countries.	81
4.9	Growth in irrigated land in South Asia and China during 1961–1997.	82
4.10	Amount of water required (in m^3) to produce one kilogram of selected food items.	85
4.11	Productivity gains from shifting to drip irrigation from conventional irrigation for selected agricultural items in India in the mid-1990s.	85

Foreword

India is increasingly being projected as an emergent economic superpower, based on the growth of its GNP. However, if the economic status of India is measured in terms of water status and the status of water rights, India is quickly becoming one of the most underdeveloped countries in the world.

There are serious concerns about the availability of freshwater, as India has 16 percent of the world's population but only 2.45 percent of the world's land resources, and only 4 percent of the freshwater resources. The per capita availability of freshwater in the country has dropped from an acceptable 5,177 cubic meters in 1951 to an alarming 1,820 cubic meters in 2001. It is estimated that freshwater availability could further decline to 1,341 cubic meters by 2025, and to 1,140 cubic meters by 2050 (see table 1.2, page 11). This is alarming, as the threshold per capita value for water stress is 1,000 cubic meters. Total water availability is 1,122 billion cubic meters.

In a World Water Development report, India ranked 120 out of 122 countries when ranked for water quality and for the ability and commitment to improving water quality. In terms of water availability, India has not fared well. India is ranked a lowly 133rd in a list of 180 countries. India's neighbours—Bangladesh, Sri Lanka, Nepal, and Pakistan—have fared better than India. They occupy the fortieth, sixty-fourth, seventy-eighth, and eightieth slots, respectively.

The deepening water crisis is a result of having treated water as an externality. It was an externality of the Green Revolution, which led to a tenfold increase in water use for the same crop production, thus decreasing water-use efficiency by a factor of ten. This water waste was never part of the definition of productivity and efficiency of industrial agriculture.

India's water crisis is rooted in neglecting India's ecology, hydrology, and water heritage. India's water crisis can only be addressed by rejuvenating our water culture and creating water democracy. It is not the stock markets that will decide whether Indian civilization lives or dies. The future of this ancient civilization will be determined by whether our rivers and water systems live or die.

Binayak Ray has made a timely contribution by focusing on India's looming water crisis. I congratulate him and thank him.

—Dr. Vandana Shiva, 2007

Preface

The availability of freshwater is increasingly becoming a major concern to all in the new millennium, as population growth, surge in developmental activities, and lifestyle changes are making increasing demands on a natural resource, water, the supply of which virtually remains static. For example, between 1970 and 2000, the per capita availability of freshwater declined on average by 40 percent. Yet it is not the scarcity of water per se that counts so much as the lack of water when it is needed most—and in some parts of the world, including South Asia, sufficient water is simply not available during the dry months. The threat of the greenhouse effect and climate change has made the situation further complicated.

Five of the South Asian countries, namely, Nepal, Bhutan, India, Bangladesh, and Pakistan, have common sources of freshwater from the three mighty Himalayan Rivers—the Ganges, Brahmaputra, and Indus—and their tributaries. The Brahmaputra and the Indus originate in the Tibetan plateau thus making China, the most populous and fastest developing country in the world, eligible to make claims on waters of these two rivers. Forty-eight percent of the Brahmaputra and about 12 percent of the Indus basins fall within the Chinese territory.

India's water policy is developed mainly through the vehicle of colonialism or post-colonial modes of Western development (Dallmayr and Devy 1998: 15–16) that totally ignored complex traditional aspects by treating them as unscientific and backward (Sengupta 1985: 1919–1938; Guha 2000: 145). Furthermore, besides India none of the South Asian countries including China in reality has a proven record of maintaining a pluralist, democratic political system that encourages public deliberation and scrutiny of decisions. Even in India, some argue that technological advances have been monopolized by an elite section and this disenchantment with the state has taken different manifestations (Gupta 1999) among the different sections of the population.

The South Asian countries have the largest concentration of population living below the poverty line in the world, so their capacity to invest in modern technology to convert unusable water to usable freshwater, and to pay for the processed water is limited.

This book argues that these countries—India, Pakistan, Bhutan, Bangladesh, Nepal, and China—have no option but to work cooperatively to ensure that available freshwater is used optimally if their development efforts are to be sustained. Unfortunately, most of these countries do not even have a sustainable national freshwater policy at present, and colonial political history has left a scar so deep that at least three of them remain antagonistic to each other, making it difficult to discuss freshwater sharing in good faith. The book addresses these issues and argues that these countries must actively consider formulating their development policies cooperatively, including their respective national freshwater policy.

A host of people have assisted in my work. I thank Drs. D. K. Dutta (University of Sydney) and Parikshit Basu (Charles Sturt University) for their strong support for this study. Thanks are due to Drs. Fereidoun Ghassemi and Graeme Byrne of the Australian National University and Professor Mark Selden of Cornell University in the United States (US) for their valuable comments on the initial draft. I thank Emeritus Professor (Rtd.) S. K. Sen of the Indian Institute of Technology, Kharagpur, India, and Dr. S. K. Acharya, former Director General, Geological Survey of India, for their valuable assistance as well, and Ms. Anjusree Bhattacharya and Ms Arpita De for drawing the maps and Mr. Eric Johns for his help with the computer work.

Professor Ben Kerkvliet of the Department of Political and Social Change at the Australian National University and the administrative staff there provided useful assistance. My colleague Dr. Ronald J. May helped me better understand complex policy issues. His readiness to engage in discussion and his illumination of complex issues has been invaluable.

I also thank Ms. Mary Gosling and her staff at the National Library of Australia for their untiring assistance. My thanks are due as well to Paul Livingston for personally rendering assistance that was beyond his normal call of duty.

I express enormous debt to the late Sunil Sen Sharma. His capacity to inspire and challenge people on water-related matters was profound. His untimely death prevents me from conveying my gratitude in person.

Last but not least, I sincerely thank Philip Grundy, OAM (Order of Australia Merit), and Tom Wells for their painstaking editing of the draft, which has improved the readability of the book enormously.

Introduction

In 2001 at a colloquium on India's tenth five-year plan I argued that flawed freshwater policy and management practices were a major policy challenge and retarded India's economic progress. I stressed that water must be given top priority in policy-planning, as the time lag between policy development and its application is long indeed (Emery and Trist 1972). This is clear from the US experience, where the level of per capita water consumption was lower in 2000 compared to the level in 1950s (Gleick 2004), and the two laws credited with this change—the Wilderness Act and the Law and the Water Conservation Fund—came into force in late 1964.[1]

I received a muted response. I was not surprised, but I expected the participants to realize the critical importance of water to India's economic prosperity.[2] Droughts and floods have been problems for centuries; the results have included suicides of farmers, inter-state water-sharing conflicts, and disputes with neighbours (Bellman 2004; Roy 2004; Sainath 1998: 260).

The World Bank and the UN have expressed their concerns about the global water crisis for some time. The Bank left no doubt about how critical the situation is, and experts have concurred (Serageldin 1995: 1–2; Vidal 1995). It is now widely accepted that the biggest constraint on the future growth of the world's economy is not a shortage of oil but of water (Chopra 1986: 97–118; Elmusa 1997: 237–38; Biswas 2001).

Rainfall in India is mostly concentrated during the monsoon months—three to four months of the year. Furthermore, more than one-third of India's freshwater comes from upper riparian countries in the Himalayan region, but few people in India understand its long-term political complexity. Also, besides cultural, social, economic, and symbolic issues, the contemporary governance concerns are human rights and equity and justice for the poor, women, and under-privileged and marginalized: "water" is intimately connected with all these issues. In many ways these are also politically explosive issues (Mehta 2000: 14-6; Djurfeldt and Lindberg 1975: 294–317).

Following the colloquium, I did a "heuristic" survey and found that while the survey participants recognized that water was an important issue, none considered it critical. A few were even unaware that harvesting and distributing water is expensive. Mark Twain's observation that "everyone talks about the weather, but nobody does anything about it" was no longer amusing to me, if I replaced "weather" with "water."

The government economic policies in the first two decades of planning (notwithstanding subsequent limited *green-revolution* success), made the entire agriculture and primary sector claim an increasingly smaller share of the GDP while continuing to be the home of a large and sometimes stagnant share of the workforce (Omvedt 1994: 332). Political parties, instead of working as agents of initiating change to uplift masses, remained active in dispersing patronage through *projects, schemes, and benefits* (Omvedt 1993: 291), and the first national water policy was not even formally developed until 1987. The second national water policy of 2002 virtually remained moribund and focused on a *supply-side solution* (D'Souza 2006: 481). Although there were some initial successes, in the medium-to-long run such policies are unsustainable. Such shortsighted policies have catastrophic consequences for the country as a whole, and the environmental and social consequences for the poor, in particular, could be disastrous. Although the seniors of the India's rural population tend to blame water crises on either continuing drought or poor karma or arrival of the *Kali Yuga*, the young, may not buy this for too long.[3]

The colloquium participants were well-off urban dwellers. They would scarcely have experienced the frustration of a slum dweller or a villager who had to make enormous efforts to obtain a bucket of clean water, or that of a farmer who did not receive water when it was most needed. Such situations can change only if stakeholders actively contribute to policy formulation, which has not been the case so far. The policy makers have ignored the basic principle that when the nature of the service permits stakeholders to organize themselves and exercise informed choice, the outcome is likely to be satisfactory to most (Batley and Larbi 2004: 222–23; Wade 1988: 210), and the *water policy* need not be an exception.

Gaping Holes in India's Freshwater Policy

Many political conflicts originate in water shortages or the desire to control water sources. The way former Bangladesh military ruler President Md. Ershad used the devastating 1988 flood to cover his failures by whipping up anti-Indian feeling is an example. When his government failed to cope with the disaster and public opinion turned against him, he did not hesitate to describe India as the main culprit and refused to accept any Indian assistance. He then attempted, not for the first time, to internationalize the water-management issue with India.[4] Many internal political conflicts and unrest also originate from an inability to access water from the riparian sources and the government's inability to resolve conflicts therein. It, therefore, becomes a critical policy concern for a multiethnic, multireligious, and multilingual India, as history shows that regionalism, provincialism, and ethnic nationalism can contribute to separatist movements (Duchacek 1986: 3–28, 91–111; Barber 1995: 165). And water-related conflicts are on the rise globally.[5]

Indian freshwater policy falls short on many counts. First, it fails to enlighten or raise the consciousness of most sections of the community on the fundamental nature of the problem that the sub-continent is the second-driest continent in the world with respect to per capita renewable freshwater (see table 1.1, page 5). Second, it hardly recognizes the critical need to have a regional focus and action plan, in collaboration with other riparian countries, to ensure the uninterrupted flow of freshwater to India. Third, there is a distinct lack of understanding at all levels of the administration that the policy issues are far more complex than people have been led to believe. And finally, the policy failed to recognize the uniqueness of India's social contexts (Lahiri-Dutt 2003:254), but more importantly the political culture of the country (Fisher 2004: 147–48).

Tragically, water management has remained single-mindedly focused on supply-side issues, with an emphasis on dam building and extracting underground water. The negative consequences have not been addressed. Two areas in particular have been neglected: recognizing water as an economic resource and encouraging user participation in policy development (Baumol and Oates 1975: 19–23; Finger and Allouche 2002: 62–104).[6] These clearly show policy makers' failure to learn from their own mistakes or mistakes of others. Worse, the focus has remained the same over the years regardless of the failure of earlier policies. There has been no shortage of evaluation initiatives, but little coordinating and learning through evaluation (Hira and Parfitt 2004: 141–42).[7]

The government's policy advisors have not looked sufficiently at innovative water management policies pursued elsewhere. A senior advisor viewed opponents of large dams with utter contempt and revealed his ignorance of how industrialized countries are approaching freshwater policy with the goal of environmental sustainability (Goel 2003). He should have known better, as Bardhan (1999) rightly argued that different disciplines of study, not looking for the same things, in the end contribute to a better understanding of the policy context. International experts now agree that a fundamental rethinking is needed about water, and the decision makers and the public at large need to think seriously about how water resources are utilized.[8]

Viewing water as a product of industrial and mechanical processes has been fundamentally called into question in the new millennium, particularly by new ecological realities (Blatter, Ingram, and Levesque 2001: 40). For example, in the United States, there has been a strong focus on decommissioning dams, as the net benefits of dam construction have been found to be far fewer than suggested. Many Scandinavian countries have virtually placed a moratorium on dam building, once considered part and parcel of their freshwater management policy.

A matter of major concern in India is the increasing level of internal conflicts between states on water-sharing arrangements. The Indian government, despite all its administrative, legal, and political muscle, has been unable to manage these conflicts between states over the last six decades. Meanwhile new conflicts have arisen as the water situation has worsened. But there has been no movement towards an integrated water policy yet.

Equally important, Indian policy makers have not provided any indication that international political issues need to be seriously taken into account. Consider:

- India is a middle riparian country with respect to the Himalayan Rivers and many of their tributaries;
- International protocols and laws on the sharing of international riverwaters, lakes, and streams are weak and still evolving; and
- Three of the water-sharing countries in the region are densely populated. Bangladesh's water-shortage problem is different from that of all the other countries. China, with about 22 percent of the world's population, has access to only about 8 percent of the global freshwater and is facing severe water shortages.[9] China owns about 48 percent of Brahmaputra basin. Pakistan, although not as densely populated as the other three countries, is already considered a water-scarce country. In view of all this, water-sharing disputes are bound to arise.

Approaches Must Change before the
Water Crisis Firmly Sets In

Of the three major rivers, the Ganges and the Indus's waters are fully utilized by the riparian countries. The Brahmaputra's water is not fully utilized yet, but both Bangladesh and India need it to augment the supply in the Ganges system during dry seasons. A little more than 48 percent of the Brahmaputra's total catchment area is in China; about 31 percent is in India; 12 percent is in Bangladesh; and 10 percent is in Bhutan (Ali, Radosevich, and Khan 1987).

China, therefore, has a natural claim on some portion of the Brahmaputra's water. Also, she has an uncanny knack for using international conflicts to her advantage, and her national interests always take precedence over regional and international issues. China's use of the American embargo on arm sales to Pakistan following the 1965 Pakistan-India war to reverse American policy towards China is an example (Bradnock 1992: 70). So is China's decision to remain an observer in the reconstituted Mekong River Commission.[10]

India's water-sharing problem with Pakistan is relatively minor, as the basic parameters were set in the World Bank–mediated 1960 Indus water-sharing agreement. However, India's decision to construct the Tulbul project on the Jhelum River has already created bad blood; the project was halted in 1987 following protests by Pakistan. It has been so controversial that Pakistan calls it by a different name, "the Wullar Barrage." Pakistan has also contested India's right to construct the Baglihar dam and the dispute has now been resolved through arbitration. Pakistan's decision to construct Diamar-Bhasha dam on the Indus River in Gilgit in Pakistan-occupied Kashmir has been vehemently objected by India.

India shares 54 rivers with Bangladesh. After Bangladesh's birth in 1971, its relations with India remained congenial for the first few years. Following the murder of its founder-president, relations deteriorated and remained shaky until the Awami League won the 1996 election. India's relations with Bangladesh are currently lukewarm. One commentator calls them worse than those that existed with Bengali Muslims in pre-independence days (Baxter 1989: 437–40 and 442–44).

India shares water from a number of rivers with both Bhutan and Nepal. Squeezed between two giant neighbours, Nepal and Bhutan maintain friendly relations with India. All Nepalese rivers are tributaries of the Ganges River system (UN 1986: 182), the lifeline of India's economy and psyche. Nepal's political future will have profound implications for India's freshwater policy (Chapman 1995; Messerli and Hofer 1995: 65–67; Thompson 1995: 106).

Equitable utilization and joint management—the central tenets of the UN-endorsed protocol on management of international rivers by the riparian countries—are the way to proceed in managing the use of international rivers, lakes, and streams (Elmusa 1997: 5). For India, a regional focus is critical to ensure that any dispute over water-sharing arrangements is kept within reasonable bounds. Often small conflicts become unmanageable and engender other conflicts. Unfortunately, given the history of mutual distrust and political disputes between riparian countries, it often doesn't take much to rekindle tensions. As South Asian countries share the same geographic, hydrographic, and climatic features, a collective approach in riparian freshwater management is no longer an option for consideration, but it is a necessity for sustainable development and future survival.

Also, the government needs to look at the internal water-sharing issues based on sound, internationally accepted principles. A failure to do so would create double jeopardy, firstly, because of the largely unknown consequences of rapidly deteriorating global environment that may require large investments at the expense of other priority needs (Gore 2006: 48, 58, 206–7, and 255; Stern 2006: i–xxvii; Perrings 2003: 2043–2059; Hansen 2006; Borenstein 2007, and Connor 2007); and secondly, internal water-related conflicts could even threaten the Indian nationhood (Kothari 1995), if the emotional-depth displayed by involved states in existing conflicts is any guide.

This book is divided into three parts: the first deals with the inadequacy of India's national water policy, which emphasizes a supply-side solution; the second, an assessment of the proposal to link the Himalayan Rivers with the Peninsular Rivers and its national and international implications. The final part considers future prospects.

Notes

1. The *New York Times* in an editorial (September 10, 2004) argued that the language of the Wilderness Act was almost spiritual in content: http://www.nytimes.com/2004/09/10/opinion/10fri3.html?th.

2. In 2005–2006, the agricultural sector's contribution (excluding fisheries and forestry) was a little more than 18 percent of the GDP at factor cost. http://mospi.gov.in/mospi nad main.htm (November 23, 2007).

3. Hindus constituted 82 percent of the Indian population in 1991 (Table 23 of the 1991 Census Report). Karma is an important concept in Hinduism and Buddhism. The *Collins English Dictionary* (Australian 1986 edition) defines it as the principle of retributive justice determining a person's state of life and the state of his reincarnations as the effect of his past deeds. According to the Hindu mythology there

are three stages (Yuga) of development in a society: It is born, it develops, and it attains maturity. After this, the society gets into a stage of annihilation (Kali Yuga), and dies due to famine, flood, etc. Then the cycle starts with the creation of a new society (Djurfeldt and Lindberg 1975: 316–17).

4. *Far Eastern Economic Review* (1988), "Stemming the flood"; October 13.

5. Gleick identified fifteen water-related conflicts during 1900 to 1950 and from 1951 to the present there were seventy-six such conflicts. There has been a gradual increase in the numbers of conflicts in each of the decades since 1950: 5 [1951–1960], 9 [1961–1970], 7 [1971–1980], 11 [1981–1990], and 44 [1991–2003]. The rate of water-related conflicts has accelerated since the 1990s (Gleick 2003a; Iyer 2002).

6. Some consider that privatization should be a part of this policy, but Briscoe (1997) argues that recent experiences showed that blind advocacy for water markets as a silver bullet that would solve all problems are not only misguided, but actually counterproductive.

7. No distinction between policy and project has been made, as both need to be evaluated to ensure that their final outcome meets their respective objectives.

8. *Yale Global Online*: http://yaleglobal.yale.edu/ (March 6, 2004).

9. Some 100 Chinese cities and towns, mostly in the northern and coastal regions, have suffered severe water shortages in recent years. It is estimated that in Beijing Central Urban District alone water demand will increase by about 38 percent, surrounding rural districts by 12 percent, and Tianjin by 36 percent before the end of this decade (State Science and Technology Commission, PRC 1991, as cited in Postel 1992: 28–37).

10. The Mekong River Commission came to existence in 1995 with four countries: Thailand, Laos, Cambodia, and Vietnam; China and Myanmar have observer status. http://www.mrcmekong.org (August 31, 2004). Also see Milton Osborne (2000: chapters 12 and 13; and 2006).

Abbreviations

ABC	Australian Broadcasting Commission
ACIAG	Arctic Climate Impact Assessment Group
ADB	Asian Development Bank
BCM	Billion Cubic Meters
BNP	Bangladesh Nationalist Party
CKM	Cubic Kilometers
CM	Cubic Meters
COP	Council of Parties
CUP	Cambridge University Press
CWC	Central Water Commission
EEA	Environment Agency
EIS	Environmental Impact Statement
EPW	*Economic and Political Weekly*
ESCAP	Economic and Social Council for Asia and Pacific
EU	European Union
FAO	Food and Agricultural Organization
FRBML	Fiscal Responsibility and Budget Management Law
GDP	Gross Domestic Product
GoI	Government of India
IAEA	International Atomic Energy Authority
ICIDI	Independent Commission on International Development Issues
ICJ	International Court of Justice

ICOLD	International Commission on Large Dams
IFRCRCS	International Federation of Red Cross and Red Crescent Societies
IIL	Institute of International Law
ILA	International Law Association
ILC	Law Commission
IMF	International Monetary Fund
IPCC	Intergovernmental Panel on Climate Change
ISWDA	Inter-State Water Disputes Act
IUCN	International Union of Conservation of Nature
IWMI	International Water Management Institute
IWRA	International Water Research Association
IWPDC	International Water, Power and Dam Construction
KM	Kilometer
KW	Kilowatt
LPCD	Liter per capita day
MAF	Million Acre Feet
MH	Million Hectares
MM	Millimeter
MW	Megawatt
NCIWRD	National Commission for Integrated Water Resource Development
NCRWC	Committee for Reviewing the Working of the Constitution
NEPA	Environment Protection Agency
NGO	Non-Government Organization
NRDC	Natural Resources Defense Council
NSS	National Sample Survey
NWDA	National Water Development Authority
NWRC	National Water Resource Council
NY	New York
OECD	Organization for Economic Cooperation and Development
OUP	Oxford University Press
PIL	Public Interest Litigation
PPM	Parts per Million
PPP	Purchasing Power Parity
SAARC	South Asian Association of Regional Cooperation
SKM	Square kilometer
SPAGC	South Pacific Applied Geoscience Commission
TI	Transparency International
TVA	Tennessee Valley Authority

UNCNR	United Nations Committee on Natural Resources
UNCHS	United Nations Center for Human Settlement
UNCSD	United Nations Commission for Sustainable Development
UNEP	United Nations Environment Program
UNESCO	United Nations Educational, Scientific and Cultural Organization
UNFCC	United Nations Framework Convention on Climate Change
UNHRC	United Nations Human Rights Convention
UNWWDR	United Nations World Water Development Report
USA	United States of America
USEPA	United States Environmental Protection Agency
WCED	World Commission on Environment and Development
WCD	World Commission on Dams
WCU	World Conservation Union
WHO	World Health Organization
WMO	World Meteorological Organization
WWF	World Wide Fund for Nature

Unless otherwise mentioned, currency units referred to this document are in US$.

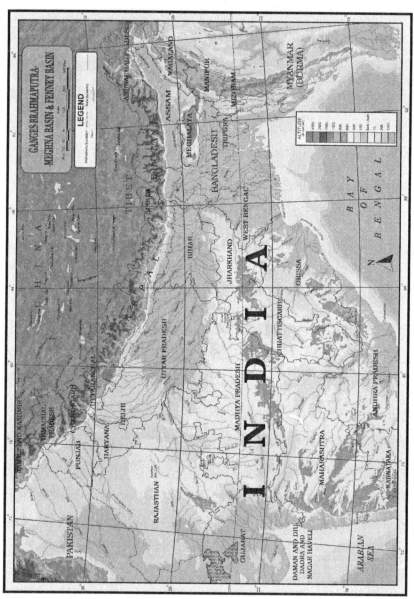

Countries sharing the sub-continental river basins, with their respective shares.

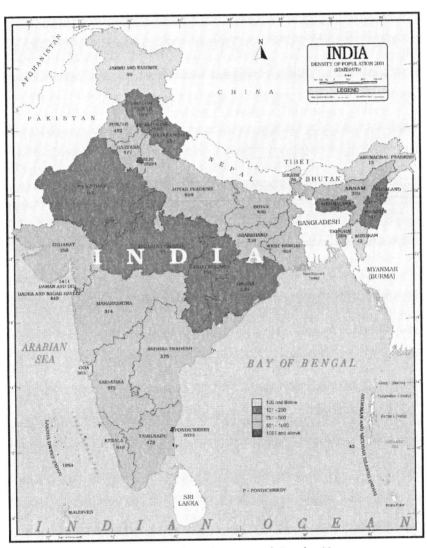

Political map of India with state-population densities.

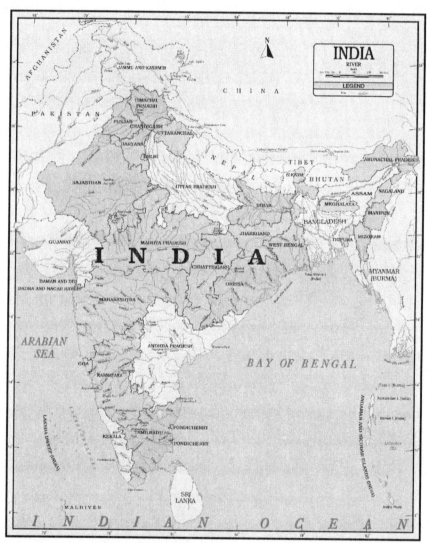

The Himalayan and Peninsular (central and southern) rivers along with major mountain ranges and international boundaries.

PART ONE

THE BROADER CONTEXT

CHAPTER ONE

The Water Environment

Global Water Environment

While two-thirds of the world's surface is covered by water, only 2.5 percent of this is freshwater suitable for human use, and it is unevenly distributed across the globe. Moreover, per capita freshwater resources are shrinking more rapidly now than in previous decades (Postel 1992: 8; Ohlsson 1995: 3; Rosegrant 1997: 1–2; UNCSD 1997; Pleming 2004).

Per capita availability sharply declined during the second half of the twentieth century in Africa, Asia, and South America. The decrease was largest in Africa—where availability of freshwater in 2000 was only 25.5 percent of the 1950 level—followed by South America at 27 percent, Asia at 34.4 percent, North and Central America at 47 percent, and Europe at 69.5 percent (Ayibotele 1992). The sharp declines in Africa, South America, and Asia were caused by rapid population growth. While birth rates in India and China (the two most populous countries on earth) are declining, elsewhere in many developing nations they remain high.

Since 1970, 1.8 billion people have been added to the planet. As a consequence, per capita water supplies worldwide are a third lower now. In the years 1970 to 2000, freshwater usage by the agricultural sector—this is responsible for about 70 percent of global consumption—increased by 175 percent. One can no longer assume that the availability of freshwater can be maintained at the existing level in all countries.

It is estimated that only an extremely small portion of the 1.36 billion cubic meters (BCM) of global freshwater is available for use. Globally each year, 500,000 cubic kilometers (CKM) are lost as moisture into the atmosphere, 86 percent from the ocean and 14 percent from the land; the continents' share in this loss is about 70,000 CKM. Small quantities of freshwater, to the tune of 40,000 CKM, are added to the earth's supply each year, products of a complex balancing of evaporation and rainfall between the sea and the land surface (L'vovitch 1979: 423–33).

In 1970 per capita water available globally was 10,000 CM but by 1995 it had declined to 7,300 CM—a decrease of 27 percent in twenty-five years. Countries become water-stressed when the per capita annual freshwater supply is between 1,000 and 2,000 cubic meters (CM). They become water-scarce when the supply falls below 1,000 CM. In 2000 the freshwater available per capita in India was 1,878 CM. Countries withdrawing between 20 to 40 percent of the available freshwater are considered medium to high water-stressed countries.

Along with hydrological and climatic factors, a host of variables affect the availability of water. For example, Cherapunji in northeast India has an annual rainfall of more than 9,000 millimeters (MM), yet a couple of months after the rains the region suffers from water shortages—a reflection of the interrelationships between water, land, and human activities, both legal and illegal (Strong and Goransson 1991; Rao 1989: 300).

Loss due to high evaporation rates is critical in tropical and arid regions. In Israel, the West Bank, Gaza, and elsewhere in the Middle East, 60 to 70 percent of all precipitation evaporates, 50 to 60 percent of it immediately (Orni and Efrat 1973: 148). Besides high evaporation, high temperature is also an important element in a country's water policy. For example, in South Asia agriculture is largely dominated by cereal crops such as rice and wheat, and rice yield is sensitive to changes in temperature: the yield declines by 15 percent for every one-degree Celsius increase in daily mean temperature.[1] The projected temperature rise in the twenty-first century, therefore, will have serious implications for India. As 65 to 90 percent of the sub-continental populations are dependent on agriculture for economic survival, issues such as the high evapotranspiration rate, rising temperature, and land salinization have profound implications for freshwater management policies.

The Situation in India Is Rapidly Becoming Critical

South Asia has the second-lowest per capita internal renewable water resources in the world, after the Middle East and North Africa (see table 1.1).

Table 1.1. Renewable Freshwater by Region

Region	Annual Internal Renewable Water Resources	
	Total in Cubic Kilometers	P/capita in Cubic Meters (1990)
Sub-Saharan Africa	3.7	7.5
East Asia and the Pacific	7.9	5.0
South Asia	4.9	4.2
Eastern Europe and former USSR	4.7	11.4
Other Areas of Europe	2.0	4.6
Middle East and North Africa	0.3	1.1
Latin America and the Caribbean	10.6	24.4
Canada and the United States	5.4	19.6
World	40.9	7.7

Source: Serageldin (1995): 2; World Bank (1992): Table A.3

Rainfall is limited to three or four months of the year in most parts of the mainland sub-continental countries (India, Bhutan, Nepal, Bangladesh, and Pakistan), which depend on the same riverine systems for freshwater. With a current population of more than one billion, India will become a water-scarce country around 2025 (World Resources 1992: 1; Postel 1992: 28–37).

India currently withdraws a little more than 26 percent of the available freshwater. Pakistan, with its rate of 70 percent, is considered a high water-stressed country, as are other countries using more than 40 percent of their available water resources (UNCSD 1997: 34 and 71). The annual Indian evapotranspiration rate varies between 1,400 and 1,800 MM. It is highest in west Rajasthan, parts of Karnataka, Andhra Pradesh, and Tamil Nadu; in some parts it often exceeds 1,800 MM. It is greater than 1,400 MM in coastal Karnataka; a narrow belt adjoining the Himalayan ranges running through Punjab to West Bengal, parts of Bihar plateau, east Uttar Pradesh, and Assam. In Bangladesh the average evapotranspiration rate is 1,250 MM, and in Pakistan it varies from 1,300 MM in northern Punjab to 2,800 MM in Sind (UN 1995: 80–81).

International Response to the Global Water Crisis

The UN, the World Bank, and the International Monetary Fund (IMF) were established at the end of the Second World War with the mandate to encourage and aid post-war economic development. The UN and Bank charters include promoting the economic and social development of developing countries, while the IMF acted as an international central bank to ensure financial stability.

Although the poorer countries suffer more than the rich countries from drought, flood, and freshwater shortages, the UN and the Bank were slow to tackle the problems of water. After dillydallying for a number of decades, they gradually awoke to the fact that deforestation, water, and other environmental questions require close attention, and realized that it can no longer be assumed that water will be an abundantly available natural resource in perpetuity.

The first focused UN initiative on water was the promulgation of the UN International Drinking Water Supply and Sanitation Decade of 1981–1990. The objective was to provide the world's population with adequate drinking water and sanitation, particularly in developing countries. This initiative was the outcome of a large number of global and regional meetings and conferences that debated water and related matters. Two conferences that directly contributed to it were the 1976 UN Habitat conference at Vancouver and the Mar del Plata conference, which adopted the 1977 Action Plan (Kinnersley 1988: 184–94; Finger and Allouche 2002: 20–61). The Mar del Plata conference is now acknowledged by most as the first world water forum (Salman 2003: 491–500; 2004: 11–19). These initiatives brought freshwater and allied questions to the forefront of the development debate and achieved some limited success. But some purported successes were actually a product of creative accounting by some developing countries that include India.[2]

The limited success led the UN to investigate other issues that contribute to the development and management of a sustainable water policy. The Rio de Janeiro conference in 1992 adopted what is now known as Agenda 21. The Agenda acknowledged the complex nature of water management: that the approach needs to be comprehensive and that all stakeholders must participate in the decision-making process.[3] The basic premises are that:

- Freshwater is finite and a vulnerable resource. Effective management of it must consider the links between land and water use across the whole of a catchment area and the groundwater aquifers;
- Water management policy should be based on a participatory approach, involving users, planners, and policy makers at all levels;
- Women play a central part in the provision, management, and safeguarding of water and should be included in all aspects of policy development; and
- Water has economic value in all its competing uses and should be recognized as an economic good.[4]

Two important reports deliberated at Rio contributed to the crystallizing of thinking on some of the water-related issues: the UN-sponsored—*Our Common Future* by the World Commission on Environment and Development—WCED (UN 1987) and *North-South: A Programme for Survival* by the Independent Commission on International Development Issues—ICIDI—1980).[5]

The UN report brought sustainability to the forefront of the global development debate, but did not assign any special importance to water, even though no substitutes exist for it. Many criticized its simplistic view of freshwater and found fault with the lack of focused deliberation on water, particularly when its availability had reached a crisis level in both Sub-Saharan and Southern Africa. One even dubbed the report *en passant* (Clarke 1991: 81). They argued that soil conservation measures should not be considered separate approaches but always in combination with soil-water-nutrient-vegetation conservation and management on the catchment basis (Falkenmark 1989: 352–53). A conference in Sweden attended by water-practitioners, academics, and the Swedish Red Cross was critical of both these reports and argued that:

- Water scarcity is a global issue, but that the burden of suffering falls unduly on people in developing and poorer countries. It acknowledged that the industrialized countries usually resort to buying their way out through their greater economic power. By contrast, developing countries are mostly trapped in poverty and debt, and have no such option;[6] and
- New approaches are needed for the proper management and use of global water resources; instead of asking how much water we need and where to get it, the global community should focus on such questions as how much water there is and how one could optimize its use (Strong and Goransson 1991).

The UN Committee on Natural Resources (UNCNR) acknowledged these criticisms in 1994 and conceded that the specter of a global water crisis had been overshadowed by other concerns, namely the ozone layer, tropical forests, and climate change. Even the Rio conference in 1992 failed to specifically address the "water crisis," and instead concentrated on "deforestation, climate change and biodiversity" (Guha 2000: 141). Many people suspect that the self-gratifying and self-glorifying objectives of large numbers of single-issue organizations shaped the report, and that they failed to consider issues other than their own.

The World Bank, in its zest to encourage economic growth, funded many multipurpose river valley projects in developing countries. In the process, and with the direct and indirect support of industrialized countries, a post-war boom in the construction of multipurpose dams took place that continued until the 1980s. Most such projects were modeled on America's Tennessee Valley Authority (TVA). Developing countries eagerly sought to emulate its apparent successes. But most project initiators either ignored or only marginally considered its environmental shortcomings and its true environmental, social, and economic costs.[7] Studies confirm that the benefit-cost analysis of the TVA was manipulated to justify its construction (Chandler 1985). Even more surprising, on the fiftieth anniversary of the building of the Hoover Dam, a U.S. senior government official publicly acknowledged that dam-irrigated agriculture was not cost-effective for a whole host of reasons (Worster 1984).

TVA's alleged poor feasibility report and the U.S. government's acknowledgment that dam-irrigated agriculture was found to be cost-ineffective are not surprising. The Colorado River is no longer the mighty river that it was. Contamination and overuse are now a serious threat to the massive Ogallala aquifer in the United States (Gleick 2004). The U.S. Environmental Protection Agency (USEPA) found that one of every three lakes in the country, and nearly one-quarter of the nation's rivers, contain pollution levels so high that people should limit or refrain from eating fish caught in them.[8]

Similar outcome is observed in India too. The Damodar Valley project was India's first multipurpose river valley project designed along the lines of the TVA. It has been unsuccessful in its stated objective of preventing floods from occurring annually, and its strong focus on the generation of hydroelectric power is misplaced—it does not produce a single kilowatt of hydro-energy now.

The Bank's overreliance on multipurpose river valley projects as a major tool for managing freshwater resources, and the failure of the projects to contribute substantially to the reduction of poverty, created doubt in many minds about the efficacy of this approach. In the process it also got entangled with the political economies of borrowing countries—which is not permitted under its charter (Rich 1994a: 8–13)—often with less than satisfactory outcome. The doubters mostly consisted of global non-government organizations (NGOs) as well as a few concerned Bank officials. The latter acknowledged that the literature on water resources was heavily influenced by the idea of comprehensive multipurpose river basin planning. However satisfying that approach might be intellectually, it was unclear whether it was the best approach to practical problems (Rogers 1992: 9–10).

This generated considerable internal political debate in the G7 countries (USA, Canada, UK, France, Germany, Italy, and Japan), forcing the Bank to acknowledge its policy shortcomings.[9] Its decisions to withdraw a funding offer for India's Sardar Sarovar Dam project and to jointly organize a workshop with the International Union of Conservation of Nature (IUCN) on large dams are examples of its acknowledgment.[10] This shift in focus is considered a watershed in the evolution of international responses to the global freshwater crisis, although a balanced water policy is still far from reality.

Outside the Bank's jurisdiction, the USSR, former Eastern Bloc countries, and oil-rich Middle Eastern countries also either followed the TVA model or undertook extremely expensive and environmentally disastrous projects to harness freshwater. These failed to set any examples that could be followed with confidence elsewhere. For example, the Karakum canal project in the USSR was considered one of the largest of its type in the world, but it is now considered to have delivered the world's least efficient irrigation system. The eventual desiccation of the Aral Sea caused by this project demonstrates the folly of such approaches. Libya's $20 billion project to transport water into the heart of the country to green the desert also failed to achieve its objectives (Pearce 1992: 3, 5, and 109).[11]

Role of Industrialized Countries

Many industrialized countries also directly contributed to the evolution of poor freshwater policies. In numerous instances their financial support for water projects in developing countries appears less than altruistic. A Nordic-funded study found that there were a number of actors operating on the political stage on the donor side, each with its own distinct role and set of interests and with mutually beneficial relationships with other actors. The study also found that, rather than delivering any benefits, the funding by industrialized countries of dam construction projects tends mostly to adversely affect peasants, tribal communities, and landless people who are already economically and socially marginalized (Usher 1997: 64; 4).[12] The actors are:

- Companies that build and supply equipment for dams;
- Aid agencies that channel grant money or credits to their recipients for the purchase of that equipment;
- Consultants who write the impact assessment on which the donors allegedly base their decisions on whether or not to finance the projects.

Given the Nordic countries' widely acknowledged commitment to global development efforts, these findings are surprising. The ulterior motivations of other donor countries for funding water resource projects would also surprise many (Goldsmith 1988: 46–47). A noted development activist argued that development assistance probably contributed more to strengthening the dualistic enclave structures of colonial economies than to advancing structural reforms towards justice, sustainability, and inclusiveness (Korten 1992: 45). It is also interesting to note that Morgenthau (1962: 264), an outstanding scholar of international politics in modern era, identified bribery as a category of aid under the bilateral aid program.[13]

Furthermore, impact of this binary approach of industrialized countries has unintended consequences for developing countries. Donor-countries' emphasis on *civil society* and *democracy* in the post-Thatcher-Reagan era openly and unabashedly forced *donor-policies* on many developing countries, particularly in the debt-ridden African countries. Consequently development funds increasingly bypass governments and go directly to charities and other NGOs which, if anything, have even less leverage vis-à-vis their sponsors than national governments: Mamdani (1990) has aptly put it, "the introduction of democracy in Africa means that Africans are allowed the privilege of deciding who is to preside over the implementation of policies designed elsewhere."

Changing Pattern of Freshwater Demand

Globally, freshwater use increased from 1,360 CKM per year in 1950 to 5,189 CKM in 2000. Asia is ahead of other continents in water withdrawals, with a rate of 61 percent in 2000. Per capita water use in a few countries such as the United States was lower in 2000 than it had been since the mid-1950s. Efficiency, conservation efforts, and technological and economic changes were responsible (Gleick 2000a, 2004).

India currently withdraws a little more than 26 percent of available freshwater resources (see table 1.2). Pakistan's withdrawal rate is extremely high; the rates for Bangladesh and Nepal are low. By 2025 India and Pakistan together will have used more than 40 percent of the freshwater available to them. But Bangladesh, Bhutan, and Nepal will have used less than 10 percent of their freshwater (UNEP 2001: 1). As India and Pakistan are likely to achieve higher economic growth rates than Bangladesh and Nepal (Asian Development Bank 2003: 281), the situation in the sub-continent will likely be of two different magnitudes: India and Pakistan will have reached their limits in using the freshwater available to them but the other three countries will have plenty.

Irrigation makes the largest claim on the world's freshwater resources, about 65 percent in the 1980s. Globally, land under irrigation has increased about fivefold during the past 100 years, from 50 million hectares (MH) to

Table 1.2. Current Annual Freshwater Withdrawals by Four South Asian Countries by Sector

Country	Annual Renewable Reserves in Billion m³	Percentage of Total Renewable Resources	Annual Freshwater Withdrawals		
			Percentage for Agriculture	Percentage for Industry	Percentage for Domestic
	Billion m³				
Bangladesh	1,217	1.2	86	2	12
India	1,908	26.2	92	3	5
Nepal	210	13.8	99	0	1
Pakistan	222	70.0	97	2	2

Note: the "Billion m³" and numbers 14.6, 500.0, 29.0, 155.6 appear under the Annual Freshwater Withdrawals section.

Source: World Development Indicators. World Bank (2003); pp. 136–37.

Table 1.3. Annual Freshwater Withdrawal by Sector and Income Group (m³)

Country Income Group	Annual Withdrawal (per Capita)	Withdrawal by Sector (Percent)		
		Agriculture	Industry	Domestic Use
Low-income countries	386	91	5	4
Middle-income countries	453	69	18	13
High-income countries	1,167	39	47	14

Source: World Bank (1992), Natural resource management in Nepal: 25 years of experience. Cited in Rosegrant (1997).

about 250 MH (Shiklomanov 1990: 34–43; Finger and Allouche 2002: 20–61). In low-income countries, 91 percent of the water is used for agriculture, followed by 5 percent for industry and 4 percent for the domestic sector. In high-income countries, 47 percent is used for industry, followed by 39 percent for agriculture and 14 percent for the domestic sector (see table 1.3).

The municipal sector accounted for less than one-tenth of global water use in the 1990s, but its share will increase significantly with urbanization and living-standard improvements (UNCHS 1991: 26–31). This is significant for India and China, as both countries are projected to achieve high economic growth rates in the coming decades, which are usually accompanied by increased urbanization. With the modernization of agriculture, a large proportion of the hidden rural unemployed move to urban areas seeking economic opportunities, which also boosts rural-urban migration. China's recent experience confirms this.

Between 1991 and 2001 the urban population in India increased from 25.7 percent to 33.4 percent of the total population, a much higher rate of increase than occurred between 1951 and 1991. Also, in 1994–1995 only 18.4 percent of households were in the middle-to-high-income group; that figure is expected to increase to 40.5 percent in 2006–2007 (Tata 2002–2003: 38; 219).

Unique Sub-continental Features

Freshwater management in the sub-continent is extremely complex because of the geo-strategic situation. South Asia is a geo-strategic region in its own right, belonging to neither of the other two geo-strategic regions of the world.[14] Despite possessing a common cultural heritage and social customs and consisting of one geographical entity, the sub-continent was partitioned in 1947, primarily on religious grounds. Religious divisions have flourished since (Chapman 2000: 3–26; 295–96), along with their negative consequences, such as intolerance and mistrust. Freshwater management policy is a victim.

During the partitioning of the sub-continent, no defensible policy was pursued with respect to ecological, environmental, and other natural resources (such as water). The restructuring of territorial lines during the transfer of power resulted in porous boundaries, with ethnic groups and communities overlapping (Phadnis 1990: 244–49; Chapman 1995: 20; Harshe 2001: 30–31).[15] Pakistan was created by carving out the northwest parts of India as a homeland for Muslims, and East Pakistan was added to it. The construction of a monolithic Pakistan with Islam as its basis failed to bind these two diverse areas together.

In pre-partition days, India, Pakistan, and Bangladesh were one political entity, but now the natural resources they use are of course no longer under one management. Political considerations and national issues became critical factors in these and their approach to other policies, though globally two hundred large rivers are shared by two or more countries and about 40 percent of the world's population depends on water from a neighbouring country (Robert 1994: 30, as cited in Ward 1997: 62–63). Furthermore, India and her neighbours have the common twin problems of endemic poverty and population rise creating enormous pressure on their limited natural resources. A peaceful sharing of water thus becomes complex many times (Chakraborty 2004: 201–8). The political imperatives of the Cold War and the mistrust created by the colonial legacy only made the situation worse.

China's annexation of Tibet and its rise as a political and economic power have further complicated the situation, as China now controls the source of two of the three major Himalayan rivers that flow through the sub-continent (the Brahmaputra and the Indus). Furthermore, with its "superpower" status, China is now in a better position to directly influence the economic and resource management policies of smaller countries in the region.

The situation in India is no better either. Within India the total disregard for ecological, environmental, and other natural considerations while establishing new states has made the situation worse within the country and has made it extremely difficult to manage the internal resources amicably, as we shall see in chapter 8.

India has two major river systems: the Himalayan and the Peninsular. The Indus, Ganges, and Brahmaputra rivers with their tributaries are part of the Himalayan system (Misra 1970: vii). These rivers have international boundaries and are shared by other mainland sub-continental countries.

Table 1.4. Percentages of (major) River Basins in Sub-continental Countries and China

Basin	Area km²	Afghanistan	Bangladesh	Bhutan	China	India	Myanmar	Nepal	Pakistan
Country area in km²		641,869	138,507	39,927	9,338,902	3,089,282	669,821	147,300	877,753
Fenny	2,800	...	34.12	...	...	65.88	...	...	...
Ganges-Brahmaputra-Meghna	1,675,700	...	7.36	2.38	19.12	62.21[b]	0.13	8.79	...
Indus	108,600	6.68	...	...	11.11[a]	26.13[c]	...	...	56.09
Irrawaddy	404,100	...	...	...	4.6	3.81[d]	91.15	...	...
Kaladan	30,500	...	...	...	...	24.24	74.39	...	...
Karnafuli	15,000	...	74.78	...	...	24.98	...	...	...

Note: a. India claims a further 0.89 percent of that basin; b. China claims a further 4.07 percent of that basin; c. China claims a further 0.15 percent of that basin; d. China claims a further 0.29 percent of that basin.
Source: Gleick (2000: 220); Wolf et al. (1999: 387-427).

India shares six river basins with seven countries in the region (see table 1.4). These are the:

- Fenny Basin with Bangladesh;
- Ganges-Brahmaputra-Meghna Basin with China, Nepal, Bangladesh, Bhutan, and Myanmar
- Indus Basin with Afghanistan, China, and Pakistan;
- Irrawaddy Basin with China and Myanmar;
- Kaladan Basin with Myanmar; and
- Karnafuli Basin with Bangladesh.

Ownership disputes between India and China relate to only small areas, but most probably what's important to them is whether the disputed areas serve strategic needs, such as defense, security, or access to natural resources such as water. Peninsular rivers do not cross national boundaries; they all originate and end within India.

With respect to the availability of freshwater, the five contiguous countries of the sub-continent—Bangladesh, Bhutan, India, Nepal, and Pakistan—fall into three groups: Nepal and Bhutan; Bangladesh; and India and Pakistan. Besides seasonal rainfall, little more than one-third of India's freshwater-flows are from outside its territory (see table 1.5). Nepal and Bhutan have a large quantity of internal freshwater resources. Bangladesh has access to abundant freshwater, mostly from seasonal rainfall and flows from other countries. Pakistan's major freshwater source is the Indus, and its water is entirely used.

At present, Nepal and Bhutan maintain cordial political relations with both India and China. China has tried to expand its ties with both Nepal and Bhutan, and in the past has even tried to undermine India's special relations with them (Dutta 1998: 101). In a rapidly changing economic and political environment, it would be rash to assume that Nepal and Bhutan will be able to maintain this balance in their relations with India and China.

Table 1.5. Freshwater Flows and Per Capita Water Available in Mainland South Asia in 2000

	Internal Flows in Billion Meters^3	Flows from Other Countries in Billion Meters^3	Total Renewable Resources per Capita in Meters^3
Nepal	198	12.0 [5.7 percent]	9,122
Bangladesh	105	1,105.6 [91.3]	9,238
India	1,261	647.2 [34.0]	1,878
Pakistan	52	170.3 [76.6]	1,610

Source: World Development Indicators (2003). World Bank; pp. 136–37.

Bangladesh receives about 10 percent of its freshwater from internal flows. Its freshwater flows are seasonal, coming during the monsoon months of the year. It suffers from water scarcity during summer months. Topography and shortage of land limit its water storage capacity (Frederiksen, Berkoff, and Barber 1993: 17).

Pakistan withdraws 70 percent of its available freshwater annually, and agriculture uses 97 percent of this water. In 1990–1991 about 82 percent of its cropland was irrigated (a higher percentage than in almost all other countries in the world). It receives 76.6 percent of its freshwater from outside the country—the Indus is the principal source of this water. It owns 57 percent of the basin. Conflicts over the share of the Indus River existed in pre-partition days, but these were all intra-country conflict. With the partition, the *share of water issue* became an issue between two sovereign nations. The 1960 Indus water-sharing agreement with India determined each country's share of the Indus water (Kirmani 1990: 200–205).

Pakistan's projected annual population growth rate of 2.2 percent during 2000–2015 is the highest in the sub-continent and twice the global projected rate of 1.1 percent. This high growth rate along with the country's high evaporation rate will undoubtedly lead to a major freshwater crisis in the not-too-distant future. There is already political tension over the use of water in Pakistan. Most prominently, people in the southern part are critical that most of the water is taken for irrigation in northern areas such as Punjab. However, water tables have also been steadily falling in Punjab, Baluchisthan, and N-W Frontier Province. People from water-scarce provinces have already started moving out of many villages; and there are indications that by 2010, Quetta, the capital of Baluchisthan, will have exhausted its supply of freshwater.[16]

China has 22 percent of the world's population but only 6 percent of its freshwater resources. Water levels in north and northeast China are dropping by about a meter a year (Postel 1992: 19; Ohlsson 1995; de Villiers 1999: 16–18). China, previously ambivalent to the climate change issue, is increasingly becoming concerned. Twenty Chinese and U.S. scientists have found that during last forty years glaciers in the western China's Qinghai-Tibet plateau have been melting by 7 percent annually, but the shrinkage has worsened since the early 1990s, and the country's chief glaciologist warned that most glaciers in the region could have melted by 2010 if global warming continued.[17] China may decide to harness water from sources that it does not presently access. That the Indian media recently published a Chinese denial that it was contemplating dam construction on the Brahmaputra River indicates that this is no longer a dormant issue.[18]

India's Freshwater Related Features

With an average rainfall of 1,170 MM per annum, India receives about 4,000 BCM of water per annum. About 1,876 BCM of this water is the average annual surface flow in rivers, between 80 to 90 percent of which takes place during four to five months of the year. During the remaining months the gross available quantity is about 400 BCM (Sen Sharma 1997: 175–82). Available yearly per capita freshwater in India is within a range of 1,000 to 2,000 CM.

India's population is projected to increase by an average of 1.3 percent per annum during the next one and a half decades. Projected growth rates in each of the sub-continental countries are higher than in India (see table 1.6). Generally speaking, the population increases in these countries will be concentrated in urban areas, as the worldwide pace will be fastest in developing countries, where the urban population is forecast to increase from 1.94 billion to 3.88 billion in the next three decades (World Bank 2002: 50).[19]

Existing socioeconomic conditions in India have significant implications for India's national freshwater policy. In 1997 the World Bank established that 44.2 percent of the populations were living on less than $1 a day and 12 percent were living were living on $2 a day (World Bank 2002: 75). This indicates that as the freshwater crisis deepens, a significant proportion of the population may not be in a position to pay for freshwater or the capacity to influence policy favourably (Omvedt 1982: 18–30).

Furthermore, as a large section of the population will not have the financial capacity to pay for water, the government will have to subsidize the water harvesting, processing, and distribution costs, possibly at the expense of other equally important public needs. Even if the states are in a position to

Table 1.6. Actual and Projected Population in the Mainland Sub-continental Countries 1980–2015

Country/Region	Population in Millions			Percentage Growth in Population 1980–2015	Average Annual Growth Rate	
	1980	2000	2015		1980–2000	2000–2015
1. Bangladesh	85.4	131.1	167.7	96.4	2.1	1.6
2. India	687.3	1,015.9	1,227.9	78.7	2.0	1.3
3. Nepal	14.6	23.0	31.1	113.0	2.3	2.0
4. Pakistan	82.7	138.1	192.8	133.1	2.6	2.2
5. Total 1–4	870.0	1,308.1	1,619.5	86.1	...	...
6. World	4,429.3	6,057.3	7,101.2	60.3	1.6	1.1

Source: Derived from Table 2.1, World Bank (2002)

cover such costs, past experience suggests that targeted clients mostly miss out. The World Bank found that the benefits of a large number of anti-poverty programs generally went to the non-poor in India (World Bank 1997: vi–vii). That could be due to poor program design and implementation, lack of focus, poor governance practices, and corruption (Ray 1999: 13). In rural areas the gentry, large landowners, and political power brokers often exercise power to obtain a disproportionate share of public goods, aided by administrative lapses and poor governance practices, which impacts on the allocation of irrigation water where a little more than 90 percent of the available freshwater is used. In such an environment, bonded and landless labourers, female farmers, and weaker sections of the community and the marginalized are most vulnerable (Reppeto 1986: 24). The freshwater policy cannot ignore these realities.

Farakka, an Issue Unlikely to Go Away

The water-sharing issue with Bangladesh remains an important element in India's national freshwater policy because, firstly, sufficient waters do not flow through the Ganges during the dry season when both India and Bangladesh need water most. The Ganges is the lifeline for a very large section of the Indian population and it occupies a unique place in the Indian psyche. Only about 8 percent of the Ganges-Brahmaputra-Meghna basin falls within Bangladesh, which limits India's capacity to be overgenerous in water-sharing arrangements.

Sharing the water of the Ganges has remained a very thorny issue since the partition of the sub-continent in 1947, although the problem existed long before the partition. The loss of the Chittagong port to Pakistan in 1947 made it urgent for the Indian government to ensure that the navigation channel for the Kolkata (Calcutta) port was not choked with silt.

From the 1850s to partitioning, eight separate investigations were undertaken to assess the state of the Ganges in lower Bengal. Most of the committees involved were unanimous in arguing that the water flows in the lower Ganges needed to be augmented to save the Kolkata port. Dredging the channel would not produce the desired outcome without the feeder channel that supplies water to flush out the silt.[20] The dredging of the river below Kolkata has been going on since 1906 (Crow 1995: 26–54).

India and Bangladesh signed an agreement in 1996 on sharing water. They agreed that there was not enough water flowing during the dry months to meet the needs of both countries and hence that they would endeavour to increase the flow then, but they have taken different positions on how to do it, particularly with respect to diverting the unused Brahmaputra waters to the Ganges (Ramakrishnan 2004). But more importantly, even if both countries agree on the mechanism, China has to be formally brought into the negotiations because of its nearly 50 percent ownership of the basin. This will require at least a multilevel approach involving these countries, and given past experience this will not be easy.

Given the religious fervour with which East Pakistan was established and Islam's inherent conservatism, a segment of Bangladesh politicians find it easy to use the Farakka issue, among others, to maintain their power bases. Often real or imaginary issues stand in the way of tackling the water-sharing problem in good faith. The crux of the problem is that other points of contention have created distrust between political leaders which they find difficult to overcome (Crow 1995: 84; Lama 2001: 176–80).

In this rapidly changing demographic and economic scenario, the physiographic environment and common freshwater sources make it essential for sub-continental countries along with China to work more cooperatively than ever before. Any reluctance to proceed with cooperative arrangements in line with the international protocols now being promoted to improve riparian water-sharing arrangements would result in disastrous environmental degradation, political unrest, and poor economic growth.

Notes

1. Information obtained privately from the International Rice Research Institute.

2. The definition of an adequate supply of safe water at the beginning of the decade was taken to mean that a source was available within 200 meters. India considers any good quality water supply within a one-and-a-half-hour walk as adequate (Clarke 1991: Chapter one; UNDP 1990). See Mink (1994: 4) for further clarification on this issue.

3. Four basic guidelines (known as the "Dublin Principles") were included in Agenda 21 of the UN Conference on Environment and Development at Rio de Janeiro in 1992.

4. Inefficient use of water can be minimized if water is considered as an economic good. This would also facilitate integration of water-policy with other economic and social policies. For example, Lenzen and Foran (2001) in Australia have established the net amount of water required to produce different types of commodities, which makes it possible to estimate the economic value of water used in various sectors and sub-sectors.

5. These were also respectively known as Brundland and Brandt Reports. The Brandt Report in particular was considered simplistic and ignored the complex issues relating to global development cooperation programs (Hancock 1989: 188).

6. In 2000 about 41 percent of the world's population was living in low-income countries; about 44 percent was living in middle-income countries; and only 15 percent was living in high-income countries. In 2015 the respective figures are estimated to be about 44, 43, and 13 percent respectively: more and more people will be pushed to either the low-income or middle-income category when the cost of water increases. Except Maldives and Sri Lanka, all other countries in South Asia are considered low-income countries (World Bank 2002: 50).

7. An analysis of TVA's overall economic records after fifty years of its existence found "evidence does not support the widely held belief that [it] contributed substantially to the economic growth of the TV region" (Chandler 1984: 7).

8. USEPA (2004), "Waterways Contain Polluted Fish"; www.epa.gov/waterscience/fish (August 25).

9. Voting rights in the management of the Bank and the IMF are determined by the contributions made by countries; in the UN it is "one country one vote."

10. Correspondence between IUCN DG and the Bank President (Dorcey et al. 1997: 134–37).

11. Pearce (1992) estimated that in total, the man-made rivers (pipelines) will irrigate fields that could be comfortably fitted into Hertfordshire, or half of Delaware. It works out at $130,000 per hectare, making them the most expensively watered fields in the world.

12. It is widely acknowledged that although donors have various objectives in providing aid; some objectives are clearly stated and others remain covert and hidden (Wood 1986: 1–67).

13. Also, see *The Economist* (1993), "Japan ties up the Asian market"; April 24. pp. 27–28.

14. Chapman (2000) identifies the other two geo-strategic regions as: the heartland of the old world—the land empires of Czarist Russia and China; and the region of maritime trade and movement—Europe, Africa, and the Americas, where settlement is largely coastal in orientation. These two geo-strategic regions provide different possibilities for movement and trade—silk roads on the one hand, the trade winds on the other.

15. Radcliffe drew the boundaries before the independence of the country. Later he acknowledged that "I had no alternative; the time at my disposal was so short that I could not do a better job" Nayar (1975: 3). Even after fifty-eight years of the partition there are still large numbers of enclaves that are encircled by inappropriate countries. For example, Bangladesh enclaves are encircled by India and vice versa. In all, 111 enclaves are involved (*Statesman*, December 18, 2005).

16. See www.earth-policy.org/Indicators/indicator7.htm and www.Irc.nl/page.php/405 (February 10, 2004).

17. *China Daily* published an article titled "Scientists warn of shrinking Tibetan glacier" and the Australian Broadcasting Commission broadcast it on October 5, 2004. http://www.abc.net.au/news/newsitems/200410/s1213705.htm (October 6, 2004).

18. On November 15, 2003, conflicting reports appeared in the print media (*Hindu* and *Statesman*) concerning this matter. *Hindu* reported that China planned to divert the waters of rivers originating in Tibet, including the Brahmaputra. *Statesman* reported that the Chinese Foreign Ministry spokesperson remained vague about the building of a dam over the Brahmaputra River in Tibet. On November 30, 2006, the Indian Foreign Minister advised the Parliament that "the Chinese leadership has denied reports that it was building a dam to divert the Brahmaputra waters to feed the Yellow river, and both countries had agreed to set-up an expert level mechanism to exchange information on hydrological data, emergency management and other issues." If anything, these reports confirm that issues dealing with water and its usage at international rivers are far more complicated than meets the eye.

19. The most recent forecast from the UN indicates that, under a medium-fertility scenario, global population is likely to peak at about 8.9 billion in 2050, an increase of about 48.3 percent (UNEP 2001). It is assumed that as population growth rates in the sub-continent have not yet stabilized, the sub-continental population will increase at about the same rate.

20. *Statesman* (2004), "Farakka Barrage" www.thestatesman.net/page.news.php?clid=1&theme=&usrsess=1&id=58420 (October 30, 2004).

CHAPTER TWO

Water Policy

An Anatomy

Policy Context

Drainage basins and watercourses have played major roles in shaping human settlement patterns. The greatest recorded civilizations are all linked to watercourses and rivers. The Nile Valley, Mesopotamia, Harappan, and the Chinese civilizations are examples.

Humanity's failure to comprehend the complexities of the relationship between water and human activities caused the downfall of many earlier civilizations (Pearce 1992: 9–27). For example, until recently it was believed that the reason for the disappearance of the Harappan civilization (2,500 to 1,700 BC) was its invasion by the Aryans. But even before the Aryans came it had started to decay, largely because of problems with its overuse of water resources without undertaking proper drainage arrangements (Prasad, Bharti, and Kumar 1987).

Indians always believed in the "sacredness" of the Indian rivers. Even the participants in the first Sepoy Mutiny in 1824 took their vows of secrecy using the Ganges water (Bandyopadhyay 2004: 28–29). Conservatively, India has about six hundred holy water places, known as "Tirthas."[1] These are either located on river banks or nearby (Kumar 1983: 205, 320–58). The great religious festival Kumbh Mela takes place every three years, rotating between the towns of Haridwar, Allahabad, Ujjain, and Nasik.[2] A staggering 30 million people gathered in Allahabad in 1989, a city of about one million. This illustrates the significance of some rivers, particularly the Ganges, to Hindus (Chapman 1995: 16; Alter 2001: 11).

National water policy must encompass all aspects of human requirements, both direct and indirect, besides all emerging issues, which might both directly and indirectly impact on the availability of water and its long-term sustainability. The issues such as impacts of climate change, evolving technology and social and political implications of policies are a few of the critical elements. In addition, a critical need is to understand the nature of the main actors, which have both direct and indirect interests in the policy. This is important because their interests may not always be necessarily altruistic. Beside the population at large, a few of the critical actors are the bureaucracy responsible for advising the government, the marginalized in the community who are often adversely affected by government polices, and the national and riparian-country politicians. They often try to outwit each other to achieve their goals. They are continuously repositioning themselves to preserve their interests. Understanding their relationships is vital to ensure the success of water policies (Peters 1999: 141–51; Mollinga and Bolding 2004: 4–8, 311–12). But an examination of India's water policies does not generate confidence that these relationships have been analyzed and the results considered.

National Water Policies of 1987 and 2002

By executive resolution the government established the National Water Resource Council (NWRC) in 1985. The Council does not have any statutory power. It is a political body designed to deal with wider policy issues so as to cushion any political opposition to broad policies. The Council includes the prime minister (chair), the Union minister for water resources (vice-chair), all state chief ministers, and several other related Union ministers.

The first meeting of the NWRC was held in 1985 in the shadows of the economic crisis of the 1980s and the widespread drought. The Council set up a group of ministers to formulate a draft national water policy. The draft was adopted by the Council in 1987 (Saleth and Dinar 2004: 167). The first national water policy was thus not adopted until after four decades of independence and six five-year plans—evidence of the low priority accorded to a national water policy. In the meantime a large number of major water projects (mainly dams) were implemented, discontents were allowed to flourish on inter-state water-sharing arrangements, and millions of people were forced out of dam sites. Strangely, the policy failed to consider the drought in any meaningful way, even though it was the catalyst for the development of a national water policy and drought management is an essential part of all long-range water-management plans (Whipple 1994: 179).

The 1987 policy had some useful aspects, but it totally avoided the constitutional issues, presumably because of divergence in views among the participants and reluctance by some states to visit contentious issues.

Besides the usual considerations such as environment, climate, and ecology, there are five other major elements in a sustainable freshwater management policy: economics, allocation, accountability, costs, and financial issues.[3] Specific policy needs such as irrigation, health, aid to people displaced by dams, and water access need to be considered within the ambit of the above management issues. Emphases vary from country to country, however, depending upon each country's circumstances.

Although the 1987 policy failed in some respects, in one area the GoI undeservedly was recognized for its achievement. The 1980–1990 UN International Drinking Water Supply and Sanitation Decade stipulated that everybody should have access to safe water within 200 meters of their residence by the end of the decade. The GoI modified the definition of access and determined that a walking distance of one and half hours was appropriate for this purpose. As a result, statistics show that the country achieved its target of providing safe drinking water to most of its citizens, a matter hotly contested by the people on the ground.

This change in definition has implications for other policies. For example, most rural women in India walk a few kilometers every day together with their children to fetch water for the family; this is done at the expense of the children's basic education (Herz, Subbarao, Habib, and Raney 1991). As they spend three hours walking, another half hour in the water queue and social conversation, and more time helping the family with such domestic and household activities as procuring wood for cooking and looking after domestic animals, how can children, particularly girls, find the time or energy to attend school?

It is little wonder that the female illiteracy rate in India is worse than that in sub-Saharan Africa. In 1991, the female literacy rate in India was only 39.29 percent, and in the rural areas it was a mere 30.62 percent. The 2001 provisional census figures show a female literacy rate of 53.7 percent.[4] Even with an increase of over 13 percentage points between 1991 and 2001, the rate lags significantly behind the male rate of 75.3 percent. And women's education is vital for the acceleration of social and economic development.

And how can expectant mothers, the aged, and the infirm walk such a long distance to access safe water? In 1991 rural aged people (sixty years and over) constituted 7.6 percent of the total rural population in India, compared to 6.3 percent in urban areas. Life expectancy at birth in India has increased in recent decades. In 1950–1955 it was 38.7 years, but in 1995–2000 it was

60.3. The proportion of people 65 and over was only 3.3 percent in 1950 but it is expected to rise to 15.1 percent by 2050 (Dhar Chakraborti 2004: 137–61). Within one generation the number of elderly people will increase substantially and many will reside in rural areas where most Indians live. This will create severe problems for a substantial number of people, as increasingly households consist only of immediate members of the family rather than the extended family.

It is universally acknowledged that the best decisions are made when stakeholders are informed and participate in policy developments that relate to them (Ingram 1990: 6). Yet in spite of women's critical role in all aspects of water management, Indian policy has remained virtually silent on their involvement. It virtually guarantees that a large section of the community will remain trapped in the poverty cycle without a realistic hope of escaping, although their understanding of rural development matters is equal to that of others (Ganguly-Thukral 1996: 1500–1503). A recent study of forestry management practices in two states confirms this (Tiwary 2004: 105–7, 165).

Both water policies lacked a strategy even to verify outcomes. For example, the policy that groundwater extractions not exceed the annual rate of recharge has no enforcement mechanism specified or put in place to verify the outcome. Estimates vary, but there is general agreement that water wells in a large number of villages in Maharastra dried up as a result of uncontrolled groundwater exploitation by sugar-cane cultivators. Sugar farms occupy only 10 percent of the state's cropland but consume 50 percent of the available irrigation water in the state, and 2,000 wells in one Taluk alone became dry (Clarke 1991: 14; Chapman 1992: 35; Food and Agricultural Organization—FAO—1994, cited in Meinzen-Dick and Mendoza 1996: A25–A30).[5]

Neither of the policies has verifiable water-ownership law. The need for a national water ownership law is simple: ground water aquifers do not stop at political or administrative boundaries. Unrestricted harvesting in one area could lead to drying up of water in other areas. Under Indian law, the ownership of land carries with it ownership of groundwater, subject to regulations and controls by respective state governments, though the surface flows are under public ownership (Iyer 2003: 307).[6]

This raises the issue of private rights versus protecting common public claims. Even in the United States, where the individual's right to property is considered sacrosanct, the judiciary refused to permit private interests to divert a portion of the Passaic River in New Jersey at the beginning of the twentieth century (Postel 1992: 60–72). It is ironic that India, a country accepting democratic socialism as its philosophical foundation, has yet to deal formally and effectively with the ownership of a vital natural resource (Isaak

Table 2.1. Violation of Pollution Standards in Some Main Indian Rivers

Name of the River	Number of Monitoring Sites	1994: Percentage Violation of Pollution Standards	1999: Percentage Violation of Pollution Standards
Cauvery	19	82.4 (17)	68.4 (19)
Ganges	27	80.0 (15)	100.0 (19)
Godavari	11	100.0 (6)	100.0 (6)
Krishna	13	55.6 (9)	55.6 (9)
Mahanadi	15	86.7 (15)	73.3 (15)
Narmada	14	69.2 (13)	30.0 (10)
Sabarmati	8	100.0 (5)	100.0 (8)
Tapti	10	100.0 (7)	100.0 (9)

Source: Kathuria and Gundimeda (2002: 141).
Figures in parentheses give the number of monitoring stations on the path of the river. The data refer to the main river only, not to tributaries.

2004: 222). And in the meantime the groundwater level is falling at an unsustainable rate, causing enormous sufferings to the poor and the marginalized that do not have any other means to access freshwater.

Indian rivers are much polluted, although large sums have been spent on cleaning these rivers (see table 2.1). And yet besides spending large sums of money without any significant positive outcome, the need for a national law to control pollution of the rivers has not been considered within the national policy context.[7] Untreated industrial, municipal, and domestic discharges in most Indian rivers are creating considerable environmental and health hazards. For example, the Yamuna, a sacred river in Delhi, receives nearly 200 million liters of untreated sewage every day. Impacts of such discharges on daily lives of those who depend on these waters are enormous. One such example is the untreated affluent released by the Dandeli paper mill in the river Kali, the only source of water for the 300,000 people in the Western Ghats in the state of Karnataka, alleged to have destroyed their land, health, and the natural habitat. There are many such examples all over India. People suffer because the administration in collaboration with the economically powerful pay scant regard to rules (de Villiers 1999: 113; Chapman and Thompson 1995: 200). Of course, the effects are mostly felt by the poor, the marginalized, and the powerless.

Being a middle riparian country with respect to the three major Himalayan rivers, the issue is a double-edged sword for India. First, as an upper riparian country to Bangladesh, India could be accused of providing polluted water to Bangladesh via the Farakka, which is politically, economically, and morally unacceptable. For example, Mexico has accused the United States of providing polluted water to it. In India's case the situation is more complex because India is a middle-riparian country to China, Nepal, and Bhutan. China, in particular, could do the same thing to India, and in that event India's position would become untenable. Both the 1987 and 2002 water policies of India failed to seriously consider this issue. Failure to come to a regional understanding may leave both their people and their economies vulnerable, as was the case with few European countries following the pollution of the Rhine and Danube Rivers.

In India both perennial and monsoonal rivers flow through number of states. Disputes concerning the share of water that cannot be resolved bilaterally become the responsibility of the central government to resolve. At the time of the adoption of the Indian constitution, the total number of states was less than twenty, and, at least, some of the inter-state water disputes could have been resolved bilaterally. Even then this was not the case and many disputes were allowed to linger for ages. Many of these disputes have been running for more than five decades and often take an ugly turn, as decisions by the tribunals or courts are most of the time challenged by the states and involved actors. Actions vary from defying constitutional provisions, as has been the case with Punjab and Haryana; Tamil Nadu and Karnataka, and Tamil Nadu and Kerala (Corell and Swain 1995: 136).[8] During the last six decades number of new states have been created and currently ongoing demands for at least another twelve autonomous or separate states exist (Oommen 2005: 230–40). The implications of the constitutional arrangements are further discussed in chapter 8.

Besides rainfall and spring-fed rivers, the three perennial rivers of India, namely, the Ganges, the Brahmaputra, and the Indus are snow-fed. The Ganges is fed by the Gangotri glacier. The glacier has already receded 14 KMs during the last one hundred years and is now receding further every year. This and impacts of other environmental and climatic factors are disconcerting:

- There is no guarantee that the glacier's decline will not accelerate as a consequence of global climate change. The border between China and India remains disputed, and the glacier's retreat may complicate the issue in the long run;
- Tectonic plate movements that are pushing the Deccan plate towards the Himalayas indicate the unstable geology of the Himalayan region. Chinese scientists now believe that "the problem of geological instability and frequent landslides alone are enough to make any huge hydraulic projects difficult or even impossible."[9] This issue needs to be critically considered in the context of the government's proposed river linking project (see chapter 5). Two further critical related issues are:

- Tectonic plate movement beneath the seabed in 2004 caused the devastating tsunami. What would happen if this were to occur in the heart of India? It would have to be a major consideration in water planning because of the immense importance of the Himalayan Rivers to India's economy and society, and
- the National Aeronautics and Space Administration in the United States found that the redistribution of large masses of water by dam projects has shifted the angle of the Earth's axis by approximately 60 centimeters since 1950.[10] India has many large dams, some three thousand of them. How will they affect the movements and formation of river courses in India?
- Indian scientists believe India has been lucky so far to have only the Koyna and Bhatsa dams, which created reservoir-induced seismicity of a greater magnitude than 4.0 on the Richter scale (Gupta 1992: 4). Most dams came into existence during last few decades and their effects on tectonic-plate movement remain unknown; so are their effects on river morphology, river courses, and the environment. It is worth noting that the Koshi, which descends from the Himalayas (of Nepal) into Bihar (India), has moved about 130 KM west in the last 200 years (Chapman 1992: 28).

Key Shortcomings of the National Water Policy

Generally speaking water policies in developing countries, over the years, have failed to incorporate a number of relevant issues. The reasons are complex, however. Many have argued that inappropriate modeling of supply and demand scenarios, and other issues such as social inequalities and inappropriately defined property rights are some of such overlooked issues (Vira, Iyer, and Cassen 2004: 312–27). While this is certainly true, it is further argued that equally important are failures to incorporate the national and international political environment, and failure to consider water policies within a dynamic economic, technological, and social environment. For example, the recent push to develop bio-fuel as an alternative to fossil-fuel in response to greenhouse concerns, particularly by the United States and the EU countries, will encourage more farmers to produce raw materials for bio-fuel than to produce food products, as almost certainly it will be financially beneficial to them. This, in turn, will add additional pressure to use more fertilizer, and additional water for irrigation. Consequential effects will disproportionately fall on the disadvantaged and marginalized section of the population in the community, as the costs of food and freshwater will certainly increase.

In developed industrial economies (any) economic and technological shortcomings of initial policies can be overcome relatively easily by making new investments in technology and reorganizing administrative arrangements. In developing countries resource bottlenecks and (often) managerial ineptitude restricts such options, hence the need to ensure that the initial policy-settings cover all appropriate areas as far as practicable. Past history certainly indicates that this has not been the case so far.

The shortcomings in India's national water policies are many. Some of these are:

- Hydrological projections have pointed to an emerging global threat caused by the dwindling supply while demand has been growing (Falkenmark 1998; Vorosmarty, Green, Salisbury, and Lammers 2000: 284–88; Revenga, Brunner, Henninger, Kassem, and Payne 2000: 25–29; Rosegrant, Cai, and Cline 2002: 4–16). The policy therefore needs to look at water-related issues in their entirety. The management of aquifers, security, human rights, equity, climate change, and ecosystems are all integral parts of freshwater management strategy (Dorcey, Steiner, Acreman, and Orlando 1997: 11; Bogardi and Szollosi-Nagi 2004: 17). The importance of these issues is evident from the experiences of other water-scarce countries, such as Israel and Palestine (Bowker 1996: 134; Lowie 1999);
- Regulatory power lies with individual state governments, but very few, if any, have passed laws and implemented them. Consequently, sustainability is neglected and equity questions are flouted. This leads to policy ineptitude, which contributes to the abandonment of agricultural land because of increased waterlogging and salinity that is costing the economy heavily (Dubash 2002: 150–51; 249; Barbier 2005: 288);
- The policy did not consider water as a scarce and valuable natural resource. Its focus has been project oriented with an evangelical emphasis on the supply-side solution. It failed to recognize that supply does not alter a water cycle; that supply management is becoming more expensive and is unsustainable on its own, while globally demand management has become more imperative; and that demand and supply have to be reconciled through a judicious mix of policies such as pricing, regulation, and education (Organization for Economic Cooperation and Development 1989: 63–64; Nickum and Easter 1994: 1–9 and 187–211; Winpenny 1994: 2; Serageldin 1995: 5–7);
- Some, however, believe that with sensible policies freshwater will always be available. Regrettably, there is no universal definition of a sen-

sible water policy (Brown, Flavin, and Postel 1991: 19, 87–88). Freshwater policy needs to be intimately linked with an individual country's level of economic development and its resource endowment (Muppidi 2004: 98). What is sensible for one country may not be so for another;

- Public policy determines natural resource distribution: who gets what, when, and how. Water policy needs to be responsive to these and other wide-ranging socio-economic issues. Neither of the GoI water policies provided any lead in this area (Ingram 1990: 31–42; Cornia and Court 2001: 2–25);

- Average per capita domestic water consumption in India dropped to 31 liters daily (LPCD) in 2000 (see table 4.2) from 50 LPCD earlier, which is the internationally accepted minimum in 1990 (Gleick 1996: 83–92, and 1999: 487–505). While China's per capita consumption was the same as India's in 1990, it increased to 59 LPCD in 2000, reflecting the country's increasingly high income. The 2002 policy did not make any attempt to unearth the reasons why Indian water consumption pattern is different from this universal trend, a critical variant in water policy to estimate demand;

- The policy objective behind the 73rd and 74th Constitutional Amendments was to ensure grassroots participation in water management. This was not even mentioned in the 2002 water policy, nor was there any mention of how stakeholders, including women, could be included in policy development and debate. The policy document even failed to give credence to the Rural Development Ministry's 1994 Guidelines on Participatory Water Development;

- The policy failed to explain why river basin authorities or boards have not been established despite the River Basin Authority Act of 1956, although such boards are common in many countries. This does not necessarily mean that *river basin planning* is the ultimate objective. It could be that in future biospheric planning may be the order of the day to overcome the climate change and other atmospheric concerns;

- It was expected that the 2002 water policy would incorporate the lessons learned from the shortcomings of the 1987 policy and ensure that they were corrected. But that was not the case. The policy statement had almost a casual tone. For example, it asked states to formulate their own water policy, with an operational action plan "in a time-bound manner, say within a *two-year* period" (Government of India 2004: 435).[11] This has hardly happened. In 2002 only Tamilnadu and Orissa had a plan and Rajasthan was in the process of preparing one (Prasad and Khanna 2002). This is not surprising, as the 2002 policy document acknowledged

that many earlier projects remained incomplete due to the states' financial constraints. It could thus be argued that perhaps the states were not keen to have a plan that might haunt them politically; and
- The reality is that the government's policy advisors have failed to distance themselves from those who discuss water-related matters in simplistic or superficial terms and are unable to consider these matters in a way that incorporates water in all its forms, namely all surface water bodies, groundwater aquifers, and atmospheric elements. The current focus on supply-side solutions allows them to pursue grandiose projects while providing only lip service to other matters. Even a former GoI policy advisor dubbed the 2002 policy a patchwork quilt lacking cogency or coherence, uninformed by philosophy or vision (Iyer 2003: 67).

Such patchwork serves two purposes, neither of which satisfies long-term policy needs. First, it allows politicians, including the administration, to make a show of their achievements to the electorate, and second, it serves the interests of the dam construction lobby. Corruption in the construction and irrigation sectors involving politicians, administrators, and others is well documented by independent studies.

Gottlieb (1988: xi), in his description of how decisions are made in the global water industry, aptly summarizes the situation. He suggests that the way the industry controls water-related decisions is little known to the public or the media, as those in the industry communicate with each other in a technically dense and often inaccessible language, and yet *public* is central to water policy.

The Indian policy makers have failed to recognize that in the new millennium, the environment, water, and their effects on human survival and the global economy can no longer be considered lightly. Increasingly, people and businesses are becoming concerned about the future of the environment and its effect on the availability of freshwater and the global economic order (Pacific Institute 2004; Carlton 2004).[12]

Sustainable use of water in the twenty-first century calls for a change in water-use practices, and a totally new outlook is required to resolve new and emerging problems, a Canadian report by the Social Science Council in 1995, "Water 2020" argued forcefully (Smith 1989: 294–95). More specifically, given the imminent freshwater crisis India is likely to face in the near future, at least three issues deserve in-depth examination with the objective to provide innovative policy directions to the community at large. These are irrigation, health, and resettlement of displaced persons caused by the ill-considered earlier policies. Unfortunately, both national policies fail in this

Table 2.2. Estimated Freshwater Savings in Agricultural Sector by Investing in Water-saving Technology

	Possible Water Savings in Billion Meters[3]					
Country	35 Percent Improvement	45 Percent Improvement	55 Percent Improvement	65 Percent Improvement	75 Percent Improvement	95 Percent Improvement
Nepal	1.01	6.25	8.70	11.60	14.50	18.85
Bangladesh	1.46	2.92	4.38	5.84	7.03	9.49
India	50.00	100.00	150.00	200.00	250.00	325.00
Pakistan	15.56	31.12	46.68	62.24	77.80	101.14

Source: Derived from various tables.

regard. The other important issue is the climate change and its impact on the environment, which, in turn will have profound impact on water policy.

Water Policy and Irrigation

The agricultural sector uses more than 90 percent of the total freshwater in India (see tables 1.2 and 1.3, pages 11 and 12). In 1997–1998, 54.6 million hectares of land were irrigated in India. More than half (56.6 percent) of this land used groundwater, compared to 31.1 percent in 1970–1971 (Tata: 2002–2003: 49). Land under canal irrigation declined from 41.5 percent of the total irrigated land in 1970–1971 to 31.3 percent in 1997–1998. This has one positive outcome, as Indian canals lose 70 percent of the water they carry before it gets to the consumer.[13] The 1999 NCIWRD report stated that eventually about 50 percent of land will depend on groundwater for irrigation, but whether enough groundwater will be available is a big question, as the groundwater level is depleting rapidly. Besides groundwater availability in the summer months for agricultural purposes, contaminated groundwater and uncontrolled extraction of groundwater by the industrial establishment remain major concerns.

Given rapidly increasing demand and static supply, a major focus of water policy has to be minimizing waste and improving water productivity. The modern approach in the agricultural sector is not to apply more water but to use it more effectively with new technology and investment. As a large proportion of the water in the irrigation sector is now wasted, the increased water demand for this sector needs to be met from the water savings in this sector. This requires large investments in new technology, increased training and education for the farming community, effective land reform, integrating land use with product value, and an efficient and corruption-free administration, among other factors (Falkenmark, da Cunha, and David 1987: 94–101). It also requires stakeholder involvement at the grassroots level.

It is estimated that if a minimum of 35 percent savings in irrigation water use is achieved, a minimum of 50 BCM of freshwater could be made available to irrigate additional land and that if 95 percent savings can be achieved an additional 325 BCM of freshwater could be made available. Table 2.2 shows the importance of making additional freshwater available for the agricultural sector by investing in water-saving technology and by improving water productivity. Obviously, savings of such magnitude cannot be achieved overnight, but a beginning has to be made.

As we will see later, the water policies followed during the last four decades have not delivered the desired outcome. Complementary policies such as land, credit, and other related reforms have not been sufficiently implemented either. Thorner discussed many of these issues in 1962 (1962: 8, 15, 62–63, 203 and 222–24), and most of his findings remain valid even today. He found that land reforms that could conceivably pave the way for a period of rapid and sustainable agricultural development had not been achieved uniformly. He also found that water was unavailable when needed most, but more importantly that it was inequitably, unevenly, and irregularly distributed. While nobody would deny that some improvements have taken place, recent work by researchers and practitioners confirm that the improvements have fallen far short of expectations.

Neither of the national water policies addressed the issues in their totality. For example, Punjab is still cultivating water-guzzling rice, for which the opportunity cost of water is extremely high, particularly since the irrigation cost has more than doubled in the last few decades (Hillel 1987: 35 and 95; Rosegrant and Svendsen 1993: 13–32).[14] It also needs to be asked why productivity in the agricultural sector, even where irrigation water is available, is still well below that of the expected level. The national water policy remains silent on these issues and on why the agricultural sector has virtually remained stagnant (Thakur 2006). Restructuring water systems depends on many things—better asset-management systems, performance measurement, stakeholder involvement, transparency, and so forth—all of which need to be addressed (Wolff and Hellstein 2005: viii; Swaminathan 2004).

Water Policy and Health

Water-related diseases are one of the most common causes of deaths in developing countries. It is estimated that globally there are 250 million cases of water-related diseases every year, and that three to five million people a year die from these preventable diseases (Gleick 2004). The real number of deaths will never be known because of inadequate diagnostic facilities and poor record keeping in developing countries including India. The 2002 water policy did not adequately address this problem.

Gastrointestinal diseases are directly attributable to contaminated water and poor sanitation. Deaths and premature incapacitation of the affected population cause huge loss to the national economy in addition to the obvious human suffering. The health situation is getting worse as rising temperatures create serious problems, particularly for the poor. Malaria has reemerged as a major killer throughout the globe (Rogers and Randolph 2000: 1763–66). Another example is the spread of dengue via the mosquitoes Aedes aegypti and Aedes albopictus, which have adapted from their natural forest environment in tropical countries to the urban environment, where they breed in pots, pitchers, water cans, and so on (World Resources 1998: 26; Struck 2006).

The naturally occurring groundwater contamination that remains mostly undetected by water users in the sub-continent also remains a major health concern. But even if people were aware of it, they would have little choice but to use such water in the absence of any alternative source. Severe health problems are caused by the use of arsenic-poisoned water in parts of India and Bangladesh.

Increased toxicity of water from streams and other sources is also creating problems, particularly in rural areas. The rich nutrient that flows from the increasing use of chemical fertilizer and mutated versions of many deadly bacteria are growing concerns. Constant free-flowing water would have carried away toxic elements or at least diluted them sufficiently. Except in the monsoon months, most Indian rivers do not have enough free flow to do that. In some cases the water cannot even be used for irrigation purposes for fear that toxic impurities might enter the food chain. The worst thing is that in most instances vast rural populations are not even aware of the existence of such toxicity. Even in industrialized countries with well-established health and environment monitoring, it is difficult to maintain adequate surveillance; it would be virtually impossible where such monitoring is poor or nonexistent.

Dangerous aquatic organisms are also causing diseases such as yellow fever and the chronic, debilitating, and potentially lethal tropical disease known as bilharzia, or schistosomiasis. These diseases have been directly linked to the construction of dams and irrigation projects; studies also show that people remain sick for up to forty-two days a year (Hunter, Rey, Chu, Adekolu-John, and Mott 1993: 50–116). The World Bank has warned that there exists a strong possibility of some of these diseases gaining a foothold in India. The World Health Organization (WHO) believes that potentially this could be disastrous (2003). Others such as Morse and Berger (1992) and Wallingford (1993) confirm this (cited in McCully 1996: 43, 56).

Increased travel and trade is also spreading many water-borne diseases, just as in earlier days when cholera spread to Europe from India (World Resources 1998: 22). That this possibility is increasing is clear from a recent report from California. The West Nile virus was first detected in New York in 1999. Eighty percent of those infected do not develop symptoms, but a small proportion does and they fall victim to severe neurological diseases. Some diseases can be sourced to moribund or sterile water.

The WHO also believes that dams are becoming nutrient-enriched quickly with intensive agricultural practices and that this increases the possibility of excessive aquatic weed growth, including one type of microscopic algae. The toxicity of these organisms has only been discovered in recent years. They are potentially lethal to humans and animals (Chorus and Bartram 1999). Other diseases such as lymphatic filariasis and Japanese encephalitis are also believed to be rapidly spreading in India and in neighbouring countries either through weed-infested reservoirs or perennially irrigated rice farming (Hunter, Rey, Chu, Adekolu-John, and Mott 1993: 4–5, 26 and 30–31). Even the much publicized Indira Gandhi Canal is now reportedly contributing to the increase in stomach ailments among children and to an epidemic of skin diseases (Goldman 1994: 131, as cited in McCully 1996: 170; Blinkhorn and Smith 1995).

Water Policy and Displaced People

Central Water Commission (CWC) figures show that India had 4,291 dams in 2002, including 695 under construction. Of these, 3,600 are large dams, and 3,300 were constructed during the post-independence period (Roy 1999: 3; Iyer 2003: 124–36; Chatterjee 2004: 479). More than one half of these dams was constructed during 1971–1989 (Gopalkrishnan 2002). Only three other countries—China, the United States, and Japan—had more dams, according to International Water, Power and Dam Construction (IWPDC 1995, as cited in McCully 1996: 6).

There are four major aspects to dam construction: environmental, economic and financial, administrative, and social. Each is important for the achievement of successful outcomes. An important aspect is the plight of displaced people from the dam sites. Although millions have been displaced to date, the government's record has been so poor that nobody knows how many have been displaced so far: a World Bank expert committee report on the Sardar Sarovar Project concluded that that the resettlement and rehabilitation in India had been unsatisfactory in virtually every project with a large resettlement component (Morse and Berger 1992). Regrettably the plight of the displaced people has been made worse than before (Dias 2002: 3). Most people displaced from dam sites depend on land for their sustenance. The Tehri Dam Project, to take one example, directly displaced 80,000 people, almost all of whom had been cultivating their land for generations (Bandyopadhyay 1995: 2367–70; Roy 1999: 18–19), and they have hardly any capacity to find other kinds of work (Patel 1994). Such rehabilitation has not usually taken place, however.

More distressing, India has neither a national resettlement policy nor an enforceable legal framework for total resettlement and rehabilitation (Asthana 1996: 1468–75; Roy 1999: ix). Whatever policy it has, it is argued, is based on a premise that displacement is an integral part of the development process, which need not necessarily be the case (Sah 2002). In a review of the

World Bank projects, Cernea (1996: 1515) argued that projects should, first, identify specific disruptions caused by displacement and to mitigate the harmful effects, if displacement is unavoidable.

The discontent created by the failure to properly resettle the displaced has the potential to become explosive. A recent example is the blockade of the Rourkella Steel Plant city by the tribals from whom land was taken away some fifty years ago. The tribals, who staged an uprising armed with bows, arrows, and axes and demanded, among other things, return to its original owners the surplus land acquired earlier. Another example is the death of dozens of people by police firing in the State of West Bengal when groups of villagers were protesting against the forceful acquisition of their land for industrial development purposes.[15] These incidents confirm that displacement-related discontents can explode at any time, particularly in an unsettled sociopolitical environment.

As there are no central records of the total number of people displaced from dam sites, researchers have made their own estimates (Fernandes and Paranjpye 1997: 15). Estimates vary wildly from 20 to more than 50 million. The truth probably lies somewhere between these two extremes. One study of fifty-six large dams found that on average about 36,134 persons were displaced per dam (Singh 1997: 187). Narmada activists concluded that some 500,000 people have been displaced every year as a direct consequence of administrative land acquisition (Kothari 1996: 1476–85; Shiva 2003: 45).

Globally, about 40 million people were displaced in eight developing countries by 1990 from various dam and other development projects (Khagram 2000: 85; Guggenheim and Cernea 1993: 2 and 204).[16] Some have even been displaced three to four times (Roy 1999: 20; 2001: 61 and 64).[17] Even the World Bank's effort to assess the gravity of the consequences of forced displacement is highly questionable. In one of its funded projects that existed for ten years, the Bank appointed overseas consultants to assess the problem who had absolutely no understanding of the values, problems, and other issues of the displaced communities, a large proportion of whom were tribals (Hancock 1989: 130).[18]

The reasons for the failure to resettle the displaced people are many: failure to understand the economic, social, and anthropological characteristics of the displaced communities; poor planning, including deliberate ignoring of issues that might work against vested interests at the various levels involved in dam construction; corruption; and last but not least, apathy of officials. These problems, along with abject poverty and ever-increasing gaps between the haves and have-nots, have created an environment of mutual distrust and antagonism between the displaced people and governments. They have also contributed to growing instability in a society riddled with contradictions created by caste and religious divisions, exploitative land-ownership practices, the existence of bonded labor, and a lack of economic opportunities for a poorly trained labor force. In this environment, the distinction between local and wider political conflicts becomes blurred. Additional factors in the government's failure to rehabilitate displaced people include:

- The large number of "tribal people" and marginalized sections of the community who were displaced often did not have legally valid documents to establish their ownership of the land they had been cultivating or using for generations. Even if they had such documents, there was a high chance that the village money lenders would have repossessed them in a dubious way. Consequently, many of these people do not officially exist and cannot claim any compensation when they are displaced;
- There has often been a tendency to deliberately underestimate the number of displaced people. A worldwide examination of 14 dam projects that significantly underestimated the number of displaced people found that six of those projects were in India (McCully 1996: 84; Shiva 2003: 53).[19] These projects reveal astonishing manipulation. One official project document showed that the estimated number of people to be displaced was about 196,000. In reality, the number was about 885,450. An obvious reason for such discrepancies is an alliance of vested interests such as the project authorities, lending agencies (Wilks and Hildyard 1994: 225–29), and the dam industry (Cernea 1994); and
- Apart from corruption, sheer bureaucratic inefficiency is also responsible for the inability to properly estimate the number of displaced people.

Adding to their travails, the displaced people on occasion become the victims of poor leadership. One recent case was the displacement of 22,000 people with the raising of the Harsud dam on the Narmada. Their case went to the Supreme Court and they lost. Many victims questioned the leadership that promised to fight to the end after they were forcibly removed from their land. There were claims and counterclaims by the leadership, but the displaced people suffered most (Singh 2004).

There are other problems. Resettlement and compensation packages have not covered those whose livelihood depended on the displaced property holders. Despite constitutional mandates of equal treatment, in an overwhelming number of cases the interests of politically and economically weaker groups and individuals have been severely violated (Kothari 1996: 1476–85). Some official documents have falsely claimed that displaced people had either been resettled or fully compensated. Two such examples, among many, are found in the Srisailam Resettlement Experience Report (Goldsmith and Hildyard 1984, vol. i: 255–60) and the Report of the Independent Commission on the Sardar Sarovar Dam (Morse and Berger 1992: 132–35). Even the government's showpiece, the Indira Gandhi Canal system in Rajasthan, reveals similar problems (McCully 1996: 170).

This problem is more serious than meets the eye. About 23 percent of the Indian population consists of marginalized communities, variously known as "tribals," "dalits," "scheduled caste," or "scheduled tribe." A report of the Scheduled Castes and Tribes Commissioner noted that even though only about 8 percent of India's population is tribal, more than 40 percent of the displaced people up to 1990 were tribal.[20] In the Sardar Sarovar dam project area alone, 57.6 percent of the displaced people were tribal. Development continues to displace people at an escalating pace in India, where an estimated one half million people are annually displaced (Kothari and Harcourt 2004: 3–7).

There are many other examples. The six irrigation and hydro-electric projects in the tribal majority Jharkhand state displaced a total of 425,000 people. A study of some 110 projects found that of the estimated 1.7 million displaced people about 50 percent were tribal. In the recent past the proportion of tribals among those displaced has been on the rise (Fernandes 1991: 243–70).

The most disconcerting fact is that while the community as a whole benefits from the completion of development projects, the people who make the greatest sacrifice by having to give up everything do not get back even the minimum possessions and pride they had before the displacement (Bhaumik 2005: 145–49). One Bagri dam–displaced person, now a slum-dweller who had 12 acres of land before being displaced, said it all: "By living like animals we have lost our pride as human beings; our children would never believe that we were once thriving farmers" (Yaswant 1993; George 2001: 125–50). There are many such examples (McCully 1996: 79). Unfortunately, social attitudes often make the situation worse for the displaced, as the resettled invariably end up impoverished, demoralized, bitter, and humiliated, and often they are scorned by their erstwhile neighbours as "submerged destitutes" (Behura and Nayak 1993).

Displacement is not on the wane. One would expect that with five decades of experience, the 2002 water policy would have forthrightly tackled the problem. The problem is inherent in the planning process itself, notwithstanding the pious statements of political leaders that the displacement of people should be a last resort in project formulation after all other options have been exhausted (Ramanathan 1996: 1486–91).

It is hard to resist the conclusion that on resettlement issues people with power have failed to give importance to the human aspect. If involuntary displacement was unavoidable, the logical option would have been to consider people's total rehabilitation needs at the starting point of the project formulation, with a clearly delineated management structure ensuring that the identified needs were met. Experience shows that in most instances displaced people are left to stumble about in a morass of uncertainties (Cernea 1996: 1515–23). The displacement issue has assumed added importance in today's political environment, as the country is faced with a number of disgruntled and marginalized communities who are crying out for redress. These groups are increasingly taking matters into their own hands, often in ways that fall outside constitutional provisions.

A few individuals within the World Bank ably supported by a Bank consultant found that it failed to provide adequate resources to evaluate projects properly or used questionable management practices, which contributed to these failures (Khagram 2000: 83–112; Rich 1994: 111–13, 260–61; Mikesell 1992: 80–82). Such failures served the interests of powerful institutions such as the dam construction lobby. For example, the Bank-appointed consultant for the Kalabagh dam project in Pakistan produced a nine-volume feasibility report that contained no mention of people to be displaced, although eventually 80,000 people were officially displaced (some believe the final number was about 124,000). This surely could not have happened without the direct or indirect connivance of officials of the host country, including politicians, and of the Bank's management (Cernea 1990: 4). Proponents are generally conscious that estimates of the real scale of displacement would weaken internal political support for a project and that the funding agencies may feel reluctant to allow such projects to go ahead (Cernea 1986: 14). Such practices have sowed the seeds of large-scale discontent.

The funding agencies' failures could not have taken place without the direct connivance of the host government officials. For example, one Indian executive director at the World Bank actively thwarted the Bank's intention to make public its environmental assessment reports for projects seeking funding; he received active support from another developing country with a questionable environmental record. The Bank was not allowed to release the information publicly. The Bank, however, now recognizes that environmental assessment is no longer a one-shot affair and that project approval has to be an iterative process linked with the progress of the project itself (Rich 1994: III–47, 260–61).

The funding agencies' role is critical in large projects, as they either directly provide funds or arrange for funds or technical assistance. If funding agencies follow procedures meticulously without any hidden agenda, two outcomes are possible: a project is found unworthy of support on the ground that the feasibility or environmental impact reports are deficient, or conditional funding is approved subject to all criteria for support being met. In such circumstances, one would expect that the resettlement and rehabilitation requirements would be fully considered and adhered to.

The government's failure to deal with the resettlement issue within a total development context is clear (Sinha 1996: 1453–60). There was no discussion of rehabilitating displaced agricultural workers or farmhands on farmland of equivalent quality in any of its policy deliberations. Even in instances when respected social activists raised this issue with the government, the response was muted displeasure instead of any firm commitment to consider the issue thoroughly (Goldsmith and Hildyard 1984, vol. I: 245).

The resettlement issue has to be considered in its entirety, which would include providing capital to enable self-employment and similar initiatives. But such steps would require a total rethinking by governments on rehabilitation (Dhagamwar, Ganguly-Thukral, and Singh 1995).

The bureaucratic high-handedness in formulating freshwater policy is evident from the fact that the draft policy was referred to fifteen GoI ministries and departments but not to the people and agencies that were intimately involved or to the state governments that were responsible for implementing the policy. The government's excuse was that it would be a time-consuming exercise. This shows how *not* to assure the success of the guidelines. The government sought only the states' agreement that the draft policy would be used as a guideline and promised to consult the states before the guidelines were finally issued. The important issues of federalism and a national, uniform rehabilitation law were nowhere to be found in the draft document, despite the fact that states have the constitutional responsibility for resettlement.[21]

The draft policy was strongly criticized by a cross section of people and agencies. The major criticisms were that:

- Policy proceeded from the assumption that displacement itself can be used as an instrument of positive change towards better physical and economic conditions for the displaced, but mechanisms to achieve this objective cannot be found;
- The objectives of the draft policy will remain unfulfilled unless those working with the displaced people are involved in these consultations; and

- Financial aspects of this policy are defined narrowly, as it failed to recognize that the financial cost of rehabilitation should include its true costs, both visible and invisible.

The two acts used for acquisition of land are the Land Acquisition Act of 1894 and the Indian Forest Act of 1865. These two acts used the concepts of *public domain* (i.e., land acquisition for public purposes), implying that owners should be compensated without being worse off in economic terms, which, unfortunately did not happen in almost all cases. In 2006 the lower house of the Indian Parliament passed the Scheduled Tribes and other Traditional Forest Dwellers (Recognition of Forest Rights) Bill 2006. It recognizes and vests in Scheduled Tribes and traditional forest dwellers—who do not have records—the land in their possession and the rights to forest land and resources. Much of the land that was acquired during the construction of large number of dams in the past fall in this category (Tharakan 2002: xi–xii). The act may be too little too late for the comfort of most displaced families.

The government introduced the Land Acquisition Amendment Bill 2000 in Parliament without recognizing the resettlement and rehabilitation rights of displaced people. Even the recommendations of the much-criticized draft of the National Rehabilitation and Resettlement Policy for Displaced Persons were not considered. Voluntary organizations vehemently opposed the bill before Parliament and drafted an alternative "Land Acquisition, Rehabilitation and Resettlement Bill 2000." The government has given no indication that it is prepared to consider it. Many including NGOs have questioned the government's sincerity on this issue (Shiva 2003: 54).[22]

Displaced People and the Insurgency Movement

As indicated, failure to rehabilitate the displaced community with dignity has the potential to create social and political unrest, particularly when two large sections of the Indian community namely, the Dalits and the Tribals have a long history of economic and social subjugation from the more powerful sections of the community, and also have become victims of involuntary displacement (Deshpande 1997: 2090–91; Oza 1997: 1790–93). Tragically the situation of most of the displaced tribals, underprivileged and marginalized people in their homelands is worse than that of refugees (Chari 2003: 23). A Planning Commission report found that even the tribal-dominated and outwardly progressive states performed very poorly when it came to rehabilitation.[23]

Regrettably this has been happening at a time when UN members have ratified the basic human rights of the internally displaced people worldwide (UN 1998; Jeevankumar 2002: 45–47). Although these principles are not binding on member countries, they remain in a grey zone between law and politics. Furthermore, there is a general acceptance that over time the provision of soft law instruments can evolve into binding customary law and many believe that this has already happened with respect to the provision of the *Universal Declaration of Human Rights*. In fact, the Economic and Social Council and the General Assembly of the UN welcomed in 2003 the fact that an increasing number of states, UN agencies, and regional non-government organizations were applying them as a standard, and encouraged all relevant actors to make use of the *General Principles* when dealing with situations of internal displacement. And already a large number of multilateral organizations have welcomed the *Guiding Principles* (UN Doc. A/RES/58/177 [2003]; UN Doc. E/CN.4/RES/2003/51 [2003]) by adopting the resolution (Fisher 2005: 327–28). The Human Rights Commission wanted the central government to amend the Land Acquisition Act in such a way that the rehabilitation of displaced persons becomes an integral part of the projects (Banerjee 2005: 307), but to no avail.

The failure to rehabilitate displaced people has wider political implications for the Indian body politic. The Naxalbari uprising in the state of West Bengal in the 1960s was the first indication of a local revolt that took on the character of a class struggle. While the government put down the rebellion with all the force of the state, it was not able to eradicate the genesis of the movement. The entrenched poverty in rural areas, together with poor administrative practices that were perceived to be mostly serving the interests of the powerful groups in their local communities, provide fertile ground for the political ideology of the movement to take root in some sections of the community. A feeling of desperation in some sections of the community has led, at least some sections to conclude that a multidimensional participatory approach is required to overcome economic injustice, political oppression and inevitable cultural discrimination (Oommen 1997: 46–77), which are providing underlying support to such movements.

The GoI has officially declared eight states "Naxalite violence-affected": Bihar, Jharkhand, Orissa, Chattisgarh, Madhya Pradesh, Andhra Pradesh, and parts of Tamil Nadu and Maharastra. Uttar Pradesh, the largest Indian state, and the state of West Bengal, birthplace of the movement, were not officially included in the government list but recently these two states suffered daring attacks on the symbols of state power. Of India's 593 districts, 157 are now affected in some measure by Naxalism—102 of them have been added to the list in the past year, and India claims that they have strong links with their Nepali counterparts, reported the *Economist*.[24]

Poverty-stricken people in rural areas, exploited by the powerful and wealthy that are backed by the politically corrupt and inefficient administration, provide the support base of the movement. It is difficult to fully gauge the number of people directly involved in this movement. If Andhra Pradesh is any indicator the situation is grim, if not desperate. Officials there estimate that since 1968 about 6,000 people have been killed, including 2,500 rebels. In October 2004 rebels accepted an offer from the government to discuss their grievances, but refused to lay down their arms when they appeared from underground. Discussions broke down when the government failed to respond to the movement's charge that individuals and companies had collectively grabbed about 27,000 acres of land around the capital city alone and its demand that any investigation of this matter should only be undertaken by the nominees of peoples' organizations.[25] It was further reported that the GoI decided to create twenty-three battalions of "security forces" specially trained and equipped to fight the Naxalites and planned to introduce sociocultural programs to ensure that the people do not feel alienated.

The national water policy thus far has failed to address many critical issues. These are: first, how best to make water available to meet the needs of the entire community in an environment where its availability in sufficient quantity remains extremely doubtful. Second, the spin-off effects of the current policy in three fronts, namely, regional relations, inter-state conflicts and the internal social unrest, need to be carefully examined before policy direction is changed to overcome these difficulties.

In the appendix of this book, one example each from Bangladesh, Sri Lanka, and the Philippines is provided to highlight that the long-term effects of the failure to judiciously address the resettlement of "internally displaced persons" could be disastrous and that can even threaten the integrity of the nation.

Notes

1. Most Hindus consider holy places as "Tirthas," which they aspire to visit at least once in their lifetime.

2. Mela is a place where fairs take place and all sorts of goods are traded ranging from household goods to domestic animals.

3. These five key elements in integrated water management were endorsed at the Expert Group Meeting for the UNCSD in 1998 (UNWWDR 2003: 328).

4. See http://www.censusindia.net/pca2001.html (June 21, 2004).

5. India is the second largest sugar producer in the world after Brazil. The global sugar market is very "corrupt"; India exported an estimated 1,630 metric tons in 2002–2003; http://www.illovo.co.za/worldof sugar/; (February 7, 2004). Indian producers are not paying a fair price for water, and taxpayers are subsidizing them to compete in the international sugar market. Indian producers generated the smallest net income per unit of water (Rath and Mitra 1989: 1–129). Rath and Mitra further found that most food grains returned two to four times as much net income as sugarcane.

6. Broadly there could be four types of property rights for water (UNWWDR 2003: 375): Open Access, Common, State, and Private.

7. On February 20, 2007, a bench of the Supreme Court expressed its displeasure with the misutilization of the Ganga Action Plan funds launched in 1985 to check the pollution of the river. The Court expressed its annoyance with the response of the Ministry of Environment and Forest on the report of the Comptroller and Auditor General which had pointed out its part in the plan's implementation. The Court observed that "all these money has been eaten away. These money could have been easily used for the pours" ("Where did Ganga funds go? SC wants to know"—*Statesman*, February 21, 2007).

8. *Statesman*: "Governments fail to agree on Mullaperiyar project" (November 30, 2006); "Bangalore road closed to TN" (February 9, 2007); "Gowda reservation over Cauvery award" (February 12, 2007); "Karnataka ready for bandh" (February 12, 2007); "IT sector buys peace, empathises on Cauvery" (February 19, 2007); "TN parties for Cauvery review" (February 20, 2007); and "Cauvery: TN gets lion's share," PTI (February 5, 2007).

9. A. Chen, Princeton University and C. Chen, Peking University in a letter to the editor in *Nature*, 429/6991.

10. Harper's Index 1997 [4] 13 & 87. In the last 100 years the North Pole has tilted 32.8 feet towards the eastern United States.

11. Emphasis added.

12. "Good news for business." http://www.pacinst.org/reports/business_risks_of_water/index.htm (August 24, 2004). http://us.f405.mail.yahoo.com/ym/ShowLetter? (March 19, 2004).

13. *Economist* (2000), "Water in India: Nor any drop to drink"; August 24. pp. 49–50.

14. Ibid.

15. The Government of West Bengal wanted to acquire land in a small rural town near the port city of Haldia for establishing a large chemical-hub. The residents strongly objected to it and were engaged in activities such as digging roads to prevent government people entering the village. In March 2007 one such confrontation took place when violence broke out and police fired on the demonstrators, killing dozens of people.

16. Eight countries with estimated displaced people are: China 20 million (1950–1990); India 18.5 million (1950–1990); Thailand 130,000 (1963–1977);

Brazil 400,000 (1980–1990); Turkey 300,000 (1980–1990); and Mexico 166,000 (twelve selected dam projects 1957–1990).

17. A. Roy (1999 and 2001) claimed that the Indian Planning Commission Secretary in a meeting in New Delhi on January 21, 1999 said that he thought the number of displaced persons was in the region of 50 million of which 40 millions were displaced by dams.

18. This displacement was not directly related to "dam construction," but the principle involved in resettling the displaced is the same.

19. These six projects are: Andhra Pradesh Irrigation II; Gujrat Medium Irrigation II; Karnataka Irrigation/Upper Krishna; MP Medium Irrigation; Sardar Sarovar; and Upper Indravati.

20. Twenty-ninth Report of the Commissioner for Scheduled Casts and Tribes; GoI; cited in Chatterjee (2004: 485).

21. Only Maharastra, Karnataka, and Madhya Pradesh have declared policies on resettlement and rehabilitation (Goyal 1996: 1461–67; Singh 1997: 184).

22. Also see a large number of articles on this issue in *Economic and Political Weekly* 31/4.

23. *Statesman* (2004), "Backwardness is a backwards' story"; November 13.

24. *Economist* (2004), "The bothersome little people next door"; 6–12 November. pp. 67–68.

25. *Statesman* (October 18 and 19, 2004): "Land edges out arm on day 3" and "Andhra Whose Who in grab club."

CHAPTER THREE

International Rivers
Global Conventions, Regulations, and India

Conventions and regulations governing the use, sharing, and management of international water resources, which include rivers, streams, and lakes, fall into two groups: laws and conventions governing the navigational uses of international watercourses and those governing the management of riparian waters. While both are important, their histories are significantly different.

The first have existed for a long time. But the second, including the 1997 UN Convention, are still in the formative stage of development and the draft Convention is yet to be ratified by the required number of countries (Salman and Uprety 2002: 8–31). However, a few of the instruments incorporated in the Convention are evolving into binding customary law. For example, the International Court of Justice (ICJ) has already used some of the provisions of the draft Convention in its decision on the water-sharing dispute between Hungary and Slovakia (Salman and Chazournes 1998: 169).

The evolutionary nature of the conventions and laws governing the riparian waters, among others, is important for three reasons. First, international riparian-water law is fundamentally different from any other body of international law because water-related arrangements or disputes involve interplay between law making, conflict management, dispute avoidance, and settlement, with a view towards curtailing and accommodating competing claims of absolute freedom (Tanzi and Arcari 2001: 305). Second, mercantilists are keen to consider water as any other commodity and ignore the fact that water cannot be so considered for the simple reason that life on earth cannot be sustained without freshwater. Most scarce goods and services respond to market forces of demand and supply and to related managerial solutions, but water may not be necessarily so. Generally speaking there has to be political involvement to bring about a harmonious solution when dealing with water-related negotiations or conflicts (Connelly and Perlman 1975: 136–45).

Third, increasingly other policy goals of nations (e.g. maintaining certain environmental standards) may require individual nations to comply with certain legal standards and practices, which would have implications for riparian-freshwater policies. This requirement has been accepted following the 1941 Train Smelter decision by arbitrators on a dispute between the United States and Canada, and was embodied in the Principle 21 of the Stockholm declaration (Schwabach 2006: 33). Most recently the global debate on climate change issues is another example and is reflected in the acceptance of the *Kyoto Protocol* provisions by almost all sovereign states.

Some nations, particularly European ones at regional levels, have come to a mutual understanding to ensure that the water flows of the riparian rivers within their jurisdictions are managed appropriately for the benefit of all the riparian countries. A few of the important region-specific conventions are:

- The 1992 Helsinki Convention on the protection and use of transboundary watercourses and international lakes;
- The 1994 convention on cooperation for the sustainable development of the Danube River; and
- The 1999 Convention on the Rhine.

These conventions are the outcomes of steps taken by a number of European institutions, including the OECD, towards achieving agreement between their member countries on a cooperative approach to managing riparian waters. They have accepted region-specific agreements, which have brought them significant benefits. This is an implied acceptance of the principle that the nature of sovereignty has been changing with the blurring of the boundaries between domestic and foreign issues (Karns and Mingst 2004: 25–26). Global concern with the *greenhouse effect* and its impact on the global environment will certainly expedite this process further.

One reason for the success of European countries in establishing some governance standards, enabling them to deal with inter-state difficulties within their national and regional development contexts, is that they have successfully established and operated certain supranational institutions that have helped in approaching many problems from a wider developmental perspective. The OECD and the European Union (EU) are examples of such institutions. Collectively, their views often guide individual countries in dealing with matters that involve a number of countries. There is no such supranational or regional body in the sub-continent to manage development or resource management issues (Onta 2001), although leadership of the member countries of the South Asian Association of Regional Cooperation (SAARC) publicly have committed themselves to achieving similar objectives.

From time to time these institutions in Europe have issued directives to avoid disasters that could cause major environmental damage and economic loss in their member countries. One disaster that they wanted to avoid a repeat of was the 1986 fire at the Sandoz factory in Switzerland that caused the death of large stocks of fish in the Rhine and forced the closing of water intakes for some towns in the Netherlands, a lower riparian country (Kinnersley 1988:3). They also wanted to halt the increasing pollution in the Rhine, which was damaging France's coastal shell-fishing industry and contaminating groundwater in the Netherlands.

Outside Europe a few such agreements have been reached, but they are not as effective as the European ones. In Southeast Asia the Mekong River Commission was formed in 1995, but it is not as effective as hoped because China, the most powerful country in the basin, has not joined it, acting as an observer only. Still, the Commission helps to generate some cooperation between the member countries (Osborne 2004: 44).

The 2000 Southern African Development Community Region Protocol on shared watercourses also accepts the basic principle that waters from international rivers need to be shared and collectively managed to optimize the benefits for all riparian countries. But it has yet to achieve the extensive objectives of the European or Mekong Commission.

The 1997 UNGA Convention

The laws governing the navigational uses of international watercourses originated with the Act of the Congress of Vienna of 1815. The historical development of these navigational uses clearly reveals the desire of the western colonial powers to:

- Facilitate the conquest of new colonies;
- Facilitate their trade opportunities and access to new markets; and
- Improve and maintain speedy communication routes between the colonies and their metropolitan centers.[1]

Over the years, while the protection of sea routes has remained a major objective of all trading nations, the same cannot be said of international river-courses. The importance of rivers as transport routes has diminished over the years, though not totally disappeared. This decline in importance has been brought about by three developments over a number of years:

- The increase in population and related economic activities made it necessary for nations to use more water from the rivers, lakes, and streams within their territorial jurisdictions. This restricted the use of rivers as international highways;
- The Cold War restricted the use of rivers as international highways in Europe; and
- New technologies made possible the use of other forms of transport at competitive cost, particularly by agencies without immediate and direct access to water transport.

While there has been a gradual decline in the use of riparian rivers for international navigation, their economic importance increased over the years since demand for freshwater has been rising. Hence, riparian countries that share water from the same rivers have become increasingly concerned that the quality of the water be preserved and that the quantity also be maintained to ensure that the lower riparian countries have adequate water.

The 1997 UN resolution establishing a convention for non-navigational use of international rivers, lakes, and streams primarily drew on the earlier work of three agencies: the Institute of International Law (IIL), the International Law Association (ILA), and the International Law Commission (ILC). The resolution is heavily based on four doctrines and a large number of other resolutions adopted over the years at various conventions, conferences, and meetings. The four doctrines are:

- *The Harmone doctrine, based on the principle of territorial sovereignty.* It implied that a country was free to use water from any system that flows through its territory, in a way that best suited it. This suggested that there were no rules of international law governing international watercourses (Lipper 1967). The doctrine negated any responsibility or obligation of a country to lower riparian countries in terms of the quantity or quality of the water that flows into those countries. This principle of absolute territorial sovereignty was rejected by various international tribunals. The ICJ made this clear in 1949 (albeit in an unusual case concerning the use of waters in an international river);[2]

- *The doctrine of absolute territorial integrity.* Contrary to the above, it stipulated that upper riparian states *must* ensure that the rights of lower riparian states are preserved by allowing continuation of the natural flow of water to those states;
- *The doctrine of limited territorial sovereignty and limited territorial integrity.* It combines aspects of the two previous doctrines. This doctrine entitles every riparian state to make natural use of a stream that flows through or originates within its territory but without taking any action that could damage others. The Permanent Court of the ICJ used this principle in a decision in 1929; and
- *The doctrine of one riparian community.* It stipulates that waters of an entire river system should be managed as one economic unit. This is an idealistic approach that ignores the realities of a world where nationalism and political interests often prevail over principle.

There are other resolutions that influenced the shaping of the UN Convention. A few of the important ones are the Madrid declaration of 1911, which focused on non-navigational use; the 1921 Barcelona Convention that confirmed the principle of freedom of navigation and made it obligatory for each riparian state to refrain from all measures likely to prejudice the navigability of a waterway; the 1923 Geneva Convention that dealt with non-navigational use of international rivers, focusing on the development of hydropower by any riparian country, subject to the limits of international law; the 1972 New York resolution on flood control; the 1979 Athens resolution that dealt with the pollution of riparian waters; the Belgrade resolution of 1980 on the regulation of the flow of watercourses and the relationship between water and other natural resources; the 1982 Montreal resolution concerning water pollution in drainage basins; the 1982 Seoul resolution on procedures dealing with transboundary groundwater issues and notification processes; and the Salzburg resolution of 1997 that dealt with definitional and procedural matters (Salman and Uprety 2002: 8–31).

The 1997 UN draft convention was not voted on by all UN members. Burundi, China, and Turkey—all upper riparian countries—voted against it. Of the sub-continental countries, Bangladesh and Nepal voted in favour of it; Pakistan abstained, as did India. The convention is yet to be ratified by the required number of countries.[3]

Opponents of the convention have argued that it is vague and thus that most provisions are open to a variety of interpretations. In the event of any dispute, they note, experts' views can be tailored to fit the politics and demands of the countries in conflict (Waterbury 1997: 279–90). Supporters argue that it is practically impossible to codify all arrangements in detail in a single format that is acceptable to all. It is precisely for this reason, they argue, that anything other than a framework convention would be unacceptable to the UN membership.

It is impossible to incorporate all possible scenarios within one convention. (One might wonder how the occupying power in Iraq after the 2003 US invasion there manages the Euphrates-Tigris waters between the riparian countries, which include Turkey and Syria, who are not friendly even in the best of times.). Scheffer (2003: 842–60), for one, argues in favour of the options available within the context of the convention. And given practical necessity, the convention has at least clearly identified the basis on which a dispute must be approached. The increasing influence of transnational NGOs and international governmental organizations has to some extent decentralized political power (Blatter, Ingram, and Doughman 2001: 8–9, 12–13), and a non-prescriptive convention would do much to facilitate the solution of disputes.[4]

Non-ratification of the convention by the required number of countries led some to compare this situation with the 1921 Barcelona Convention. It reconfirmed the principle of freedom of navigation on sea routes but was only ratified by twenty states during the first fifteen years of its existence. And it has also been suggested that many states did not follow the convention's requirement. A simple answer to this is that the colonial powers that were primary beneficiaries of this arrangement had enough maritime power to successfully confront any recalcitrant country that might block such waterways, irrespective of formal ratification. It is worth remembering that the council of the League of Nations consisted of five victor powers—Britain, France, the United States, Italy, and Japan plus four rotating members—and these five victor powers were also largest naval powers at that time (Kennedy 2006: 9–11), and they would have no problem in ensuring that others follow the provisions of the Convention.

In the post–September 11 era, even this time-tested convention is being challenged. In March 2005 Australia unilaterally extended its maritime security zone by declaring that its naval ships would intercept ships 1,000 nautical miles from its shorelines to determine their destination and type of cargo to prevent any possibility of terrorist attacks via sea routes. Australia's action is not backed by any international convention or law. Indonesia and Malaysia, two of Australia's close neighbours, have strongly objected to it, and it will be interesting to see how the situation develops should any ship refuse to respond to the Australian demand.

A Major Limitation of the UN Convention

In its definition of watercourses, the UN Convention does not include transboundary groundwater, though the ILC incorporated atmospheric water into its own definition (Cano 1989: 167–71). Even before the convention was drafted, the UN Economic Commission for Europe asked that hydrodynamics research into aquifers be encouraged through international cooperation and adopted two resolutions:

- Groundwater should be recognized as a natural resource with economic and ecological value; and
- Groundwater pollution is interlinked with the pollution of other environmental media such as surface water, soils, and the atmosphere (UN 1989).

It is now possible to bring transboundary groundwater within the convention's domain, as the IAEA has established that transboundary groundwater is no longer an isolated entity but linked with other water resources (Borole, Gutpa, Krishnaswami, Datta, and Desai 1978: 181–201; Nair, Pendharkar, Navada, and Rao 1978: 803–26). The technology has been applied in a number of countries and has achieved the expected outcome. The research has progressed further since the initial success.[5]

The effects of bilateral water-sharing agreements signed by India with riparian countries on the aquifers of third countries are not yet known. Should a third country ever feel that these agreements have adversely affected its ground aquifers, and should such concerns be supported by technical findings, it is conceivable that it may ask the other parties to revisit the agreements in question. The basis for challenging them could be that the agreements adversely affected the quality and quantity of groundwater, which substantially depends on the flow of transboundary water (Blomquist and Ingram 2003: 162–69).

In such a case the countries may be forced to agree to an assessment by independent agencies of the impact of the agreements on aquifers. Adverse findings could result in high costs, including the payment of compensation to the adversely affected country and loss of face. This is not an unimaginable scenario, as, ecologically and hydrographically, eastern India, Nepal, Bhutan, and Bangladesh are indivisible.

Riparian Countries and India's Water-sharing Arrangements

A critical problem for India is that most rivers and streams in East and Northeast India crisscross a number of countries. Consequently, international conventions and legal instruments appear most of the time to be inadequate for resolving conflicts. But, more importantly, notwithstanding the 1997 draft UN Convention, the international law on transnational river-basins is in a weaker position, as there are neither established institutions nor established doctrines that could successfully resolve inter-country conflicts on common river basins (Frey 1993: 54–68). The problem is further compounded by three additional factors: one real, one perceived, and one practical:

- The region has the largest concentration of the world's poorest people, with a high rural population density. Timely availability of freshwater is vital for their economic survival. This enables politicians to use water to gain political mileage in their home bases, at the expense of the long-term interest of their country;
- India is often treated with suspicion by the smaller countries in the region, which on occasion hinders the resolution of either the perceived or the real causes of discontent (Bradnock 1992); and
- As in an organization, the greater the number of players (and often coalitions of players) the more multiple the interests, and hierarchies are in constant flux. These complicate the processes of multilateral diplomacy and negotiation—of finding common ground for reaching agreement on collective action, norms, or rules (Karns and Mingst 2004: 27).

India has signed five treaties on water-sharing with Bangladesh, Bhutan, Nepal, and Pakistan. These are:

- The Farakka Treaty that was signed with Bangladesh in 1996 on sharing the Ganges's water. One difficulty is that the Ganges no longer reaches the sea: the upstream diversions and other water demands do not leave enough water for the river to reach its natural outlet in the Bay of Bengal (Biswas: 2001: xii). The Treaty is only a partial agreement, as the problem of augmenting the supply of water in the dry season remains unsolved. Major differences exist on how to augment the supply in the dry season;
- India has signed three treaties with Nepal: the Kosi, Gandaki, and Mahakali. These focus on matters that relate to India and Nepal only. In the long run this could create problems if it can be proven that the agreements are affecting lower riparian countries, either directly or indirectly; and
- The agreement on the Indus water that was signed in 1960 with Pakistan (the World Bank being its midwife). It provided for equitable ap-

portionment of water from the Indus system and put in place institutional arrangements for the implementation and management of the treaty (Caponera 1987). As the sharing arrangement was not based on an integrated approach, problems could occur in the future.

Each of these treaties, it can be argued, is pregnant with huge environmental, economic, and political problems, either because the treaties need to involve third countries during their negotiation stages or because their outcomes would have a detrimental impact on another riparian country. And herein lies the importance of a regional approach in freshwater management policy. That regional approach may require incorporating biospheric or river-basin planning on a future date depending upon individual circumstances.

Although India shares fifty-four rivers and streams with Bangladesh, the Farakka Treaty only relates to sharing the Ganges's water. It will remain in force for thirty years and can be renewed with mutual consent. The Farakka barrage was initially proposed in the second half of the nineteenth century to augment water flow in the lower stream of the Ganges so as to maintain navigability for ships entering Kolkata port. Since its construction after independence, it has overshadowed other disputes between India and Bangladesh.

Water sharing between India and Bangladesh during the dry months remains a major concern because of the decreasing flow of water from the upper stream. Many on both sides believe that a wide gap exists between the positions of the two countries on this issue; the matter will continue to provoke debate until the flow can be augmented in the dry season (Iyer 2003: 236; Ray 1998; Crow 1995:15; 75). The sharing of the water of some of the other fifty-three rivers—such as the Teesta in North Bengal—and the Indian decision to proceed with the Tipaimukh dam in Northeast India also remain potentially contentious.

India and China share the Brahmaputra basin, with China having 48 percent of the basin's share. They have already fought one war over a boundary issue, and although four decades have passed since that conflict the boundary has yet to be demarcated to their mutual satisfaction. Sharing Brahmaputra water has not yet become a contentious issue yet, but there is no guarantee that future consideration of it can be organized without rancour.

China's decision to remain an observer in the Mekong River Commission reflects its policy of putting its national interest first. As an observer, it has not had to take any responsibilities; at the same time it is free to reap the benefits of the Mekong and keep an eye on the activities of the other countries in the basin. China has already constructed dams on the upper stretches of the river which are causing economic hardship to the lower riparian economies in particular (Osborne 2004: 17–23).

Nepal and Bhutan are squeezed between India and China. Both countries officially have good relations with India and China. Nepal resolved its border problems with China in 1961, although India denounced the agreement (Chung 2004: 104). For a number of reasons, some justifiable and others not, a significant portion of the Nepalese community feels antagonistic towards India. The new and emerging political structure in Nepal has already led to some political division, as former ministers belonging to the *Madhesi* community in the Terai region organized demonstrations arguing that "the Interim Constitution has 'failed' to address the issues raised by the Terai people." And the Terai people in many respects are closer to India, economically, ethnically, and culturally. These internal problems have a tendency to significantly influence the foreign-policy approach in the sub-continental countries. Nepal being an upper riparian country to India, future prognosis on any water-sharing arrangements thus remains uncertain.

The three treaties India signed with Nepal—the Koshi (1954), Gandaki (1959), and Mahakali (1996)—have caused serious strains, although subsequent amendments to the Koshi and Gandaki treaties eased some Nepalese concerns (Iyer 2003: 223). The Mahakali treaty, which can be reviewed every ten years or even earlier, is focused on "the integrated development of the Mahakali River and covers the Sarada and the Tanakpur Barrages, and the proposed Pancheswar project." The treaty establishes a bi-national Mahakali River Commission guided by the principles of equity, mutual benefit, and no harm to either party; it was approved by the Nepalese Parliament. A number of social and environmental activists in both countries came down heavily on both governments concerning some of the provisions of this treaty.

India's relation with Nepal has remained in the grey area for some time. The requirement in the 1990 Nepalese constitution that all treaties with foreign countries involving the use of Nepalese natural resources or affecting the country "in a pervasively grave manner or on a long-term basis" be approved by a two-thirds majority of Parliament was directed to India. Some obstructions to any agreement are bound to be ill-conceived, because once mistrust takes root, proposals are judged by appearance rather than substance. Hence India has to tread carefully in all negotiations with Nepal and show sensitivity and understanding (Iyer 2003: 223–31).

There are other sensitive issues too. Nepal is a landlocked country, and questions of transit, border demarcation, and the balance of trade remain contentious. But these are political sleeper issues to most of the endemic poverty-stricken population. In Nepal about 42 percent of the population lived below the national poverty line in 1995–1996 and 82.5 percent was living on less than $2 a day (World Bank 2002: 69).[6] As with the very poor elsewhere in the world, it is not difficult to divert their attention from anything that does not immediately affect their day-to-day lives.

Bhutan has friendly relations with India. The former king's abdication of the throne in favour of his son and the establishment of a constitutional monarchy usher in a new era in the relationship between these two countries. Immediately after the accession to the throne the new king Jigme Khesar Namgyel Wangchuk began to update the India-Bhutan friendship treaty of 1949 which the GoI has hailed as a historic moment in its relations with Bhutan.[7] It is a good beginning, but given the realities of international politics, and given the underlying rivalry between India and China, it remains within the realm of possibility that Nepal or Bhutan may play China off against India (Chapman 1995: 20; Bakshi 2004: 274–75).

Pakistan, whose economy is dominated by agriculture, relies on irrigation water for 90 percent of its agricultural needs. The 1960 Indus Treaty between Pakistan and India has been operating reasonably well, with occasional hiccups. The dispute over the Salal hydroelectric project was resolved, but those over the Tulbul and Baglihar projects were not. The Tulbul project was put on hold. Pakistan's animosity towards this project is so intense that it does not want to acknowledge it by the Indian name, and has given it a new name, Wular barrage (Iyer 2003: 220–21). The Baglihar project was referred to arbitration, and the arbitrator's recommendation has been accepted by both countries.[8] An independent expert's recommendations virtually endorsed India's original position and suggested only a small height-reduction of the dam structure. However, earlier the Pakistan prime minister publicly acknowledged that disagreement with India on the Baglihar dam construction had affected Pakistan's recent talks with India to normalize relationships; at the same meeting, he had the warmest of words for China.[9] This clearly shows that *water-related* disputes can adversely impact the wider relationship between the disputing countries.

Pakistan is a heavily indebted country, with a net debt of 45 percent of its gross national income, compared to India's 16 percent (World Bank 2002: 269). Given that nearly 85 percent of Pakistan's population is living below the international poverty line, the government may not be in a position to make large investments to increase freshwater supply, as the water crisis deepens. Pakistan, therefore, would probably be tempted to revisit the existing water treaties, and it might even keep the disputes alive for internal political reasons.

Pakistan's alleged involvement with the Kashmir insurgency movement (Gossman 2002) and the three wars it has fought with India have not helped develop a bilateral relationship that would contribute to resolving water-sharing issues in good faith. Strong international pressure to fight terrorism has fostered some easing of tension, but it would be wrong to assume that any renegotiation of water-sharing arrangements will be easy.

Pakistan has voted in favour of the 1997 UN Convention but has expressed reservations about a number of articles. Because of its reservations, water disputes with India might have to be resolved by arbitration. Although the arbitration process does not necessarily carry predetermined baggage, it is not value-neutral in absolute terms. Arbitration does carry with it a large element of uncertainty, particularly in a world where political considerations often take precedence over strictly technical ones.

Agreements Can Be Questioned, Renegotiated, or Rescinded!

In colonial days, agreements were mostly forced upon colonies by their colonial masters. In post-colonial days other political imperatives encourage governments to denounce agreements reached by their predecessors, sometimes justifiably, but often merely to secure political advantage. Furthermore, in a rapidly changing political and economic environment, demands for renegotiations of agreements become intensive given development needs and governments' efforts to demonstrate their political and economic credentials to their people.

Kenya's notice of its intention to withdraw from the 1929 Nile Basin Treaty provides an example of how allegedly disadvantaged riparian countries can opt out of agreements signed by former governments. There are ten countries in the Nile basin, and Egypt, the one furthest downstream, has a guaranteed flow of 55 BCM of Nile water a year under the treaty. Since gaining independence, poorer countries of the upper basin have questioned the terms of the 1929 agreement,[10] although without the waters of the Nile, Egypt will become a desert: 95 percent of the country's hydraulic resources come from the Nile, but only 28 percent of the population lives on the banks of the river and consumes 66 percent of the water (Nkrumah 2004).

The political environments in three of the four riparian countries that India has reached water-sharing agreements with—Bangladesh, Nepal, and Pakistan—remain uncertain. Despite the agreements, the freshwater management policies of these countries need to be constantly monitored and points of contention identified long before they emerge in the public arena, as domestic issues are in reality what determines a country's approach to international issues (Putnam 1988: 427–60).

In sharing water with other countries, one of the biggest dangers lower riparian countries face is the pollution caused by operational lapses or accidental disasters in upper riparian countries. It can affect the lower riparian countries disastrously by damaging the basin environment, threatening public health, causing loss of valuable marine life and property, and displacing people. The Sandoz Blaze in 1986 in the upstream Rhine and the slippage of cyanide from the storage dams in Romanian gold mines in 2002 are examples.

India needs to be always on guard to ensure that the natural flows of riparian rivers are maintained by the upper riparian countries and that legal and administrative arrangements are in place to enable her to take appropriate steps if something goes wrong with these rivers and streams.

In the ancient times emperors and rulers accepted the principle that clean freshwater be made available to their subjects. Romans accepted it: the well-developed code of law expounded by Emperor Justinian in the sixth century declared that there could not be ownership of flowing water. Even the early English laws had similar provisions. Indeed, the Magna Carta included a provision that all permanent *fish-weirs* be removed and excluded from the River Thames to give priority to navigation. The thirteenth-century Spanish laws included enlightened water laws similar to those of the Romans (Kinnersley 1988: 36–38). The modern world now accepts the principle that access to clean water is a fundamental right. The UN formally endorsed this principle in 1998. It has gained additional importance with the US Senate move to make it an important US foreign policy plank (Stephens and Bullock 2004).[11]

Put simply, India's problems are twofold. As a lower riparian country it needs to keep a close watch on the water management policies of upper riparian countries, and at the same time it is obliged to ensure that lower riparian countries also receive what they're entitled to. The desired policy outcome can thus only be achieved when all the countries collectively develop a mechanism to manage and share the riparian waters. The best way to achieve this objective is to consider the issues from a regional perspective, as each country will then have a stake in whatever arrangements and procedures are adopted. Only a regional approach can ensure sustainability.

Notes

1. There are many such instances in colonial history. For example, the Sepoy Mutiny in 1857 led to the speedy development of India's railway network.

2. Immediately after the end of the war a British warship was hit by a mine while sailing through territorial waters of Albania. Although mining of waterways can be considered a part of the battle strategy, the Court made Albania responsible for its failure to keep the waterways safe. The principle involved in this judgment was reconfirmed by a decision of the Arbitration Tribunal in 1957.

3. The Convention definition of a watercourse includes only groundwater connected to surface water, and does not extend to confined groundwater. The ILC later confirmed this (Salman and Uprety 2002: 27).

4. *NY Times* editorial, September 27, 2004.

5. See IAEA 2003 Annual Report (Published in 2004): B.1.4 "Using Isotope Techniques to Study Water Resources."

6. Adjusted for purchasing power parity.

7. "PM, Bhutan king ink updated friendship treaty": a statement issued by the External Affairs Ministry spokesperson, New Delhi, February 8, 2007.

8. *Statesman* (2007), "Expert clears Baglihar for reduced dam height," February 13.

9. *Statesman* (2005), www.thestatesman.net/page.news.php?clid=1&theme=&usrsess=1&id=66973 (January 29).

10. The 10 Nile basin nations are: Burundi, the Democratic Republic of Congo, Egypt, Eritrea, Ethiopia, Kenya, Rwanda, Sudan, Tanzania, and Uganda. Collectively, these countries are among the world's poorest and least developed, and almost all have been ravaged by conflict; see *Liberation* of France [2/27, 2004]; and *Yale Global Online*; http://yaleglobal.yale.edu/ , March 6, 2004.

11. Voice of America News (2005), March 17.

CHAPTER FOUR

Future Demand for Water and Available Options

Coleridge's famous line "Water, water, everywhere, nor any a drop to drink," penned more than two hundred years ago, no longer represents only the words of a frustrated mariner.[1] Alarm bells are ringing across countries, with ever increasing demand for freshwater caused by population growth and increased demands for freshwater because of expansion of irrigated agriculture, industrial development (Gleick 1998: 54), and lifestyle changes caused by increased affluence (e.g., recreational purposes).

However, in spite of unprecedented global economic and technological growth, over a billion people in the developing world still lack access to safe drinking water. An estimated 14,000 to 30,000 people, mostly young children and the elderly, die every day from water-related diseases, although nearly three decades have passed since the 1977 Mar del Plata resolution that promised people—irrespective of their social and economic conditions—the right to safe drinking water.

Assessing the water needs of a country is different from assessing needs for other natural resources. A common mistake is assessing the need as a static phenomenon (Ohlsson 1995: 15). This is a mistake for several reasons. Demand is basically influenced by a nation's level of economic development. For example, annual per capita water withdrawal in high-income countries is three times higher than in low-income countries: 1,167 to 386 CM. But while demand in the agricultural sector declines with rising income, it increases substantially in the industrial, domestic, and municipal and recreational sectors. On average, high-income countries' consumption patterns are different from those of low-income countries in use-purpose: in agriculture it was 39 to 91 CM, in industry 47 to 5 CM, and in the domestic sector 14 to 4 CM. (see table 1.3, page 12).

Water Demand for the Future

We have seen in chapters one and two that during the second half of the twentieth century global water consumption has increased dramatically, primarily due to the increase in the population growth rate and the need to produce more food. And at the same time a significant proportion of the global population still does not have access to clean freshwater, notwithstanding the development programs pursued by the multilateral agencies such as the UN and the World Bank. One thing, however, is certain: that demand for water will continue to rise, whereas the natural supply of freshwater will virtually remain static.

While all living things including flora and fauna require water for survival, demand for water is also income-elastic. Within this context it is extremely hazardous to make an estimate of future demand in a given society and at a particular point in time. Population number is the basic element that determines the basic amount of freshwater required (e.g., food production and preparation, cleaning and other hygienic purposes). Additional needs are determined by the level of other activities—both economic (e.g., industrial activities) and non-economic activities (e.g., recreational demand). As a rule of thumb, generally speaking, future demand for water is estimated on an average per capita basis. This, of course, needs to be adjusted when specific water-guzzling activities are planned or programmed that require above-average levels of water consumption. For example, in the agricultural sector cotton cultivation requires well above the average level of water. Similarly, industries such as paper or cement require well above the average level of water per unit of production.

The Food and Agricultural Organization (FAO) has made comprehensive estimates for water demand at country levels.[2] It suggests a coefficient of 1,570 CM per head of population based on a mixed animal-and-plant-based diet of 2,700 daily calories per person. It argues that this level meets nutritional standards.[3] However, as the standard of living improves, people change their food habits and lifestyle, with enormous implications for water demand. For example, the typical diet of a meat-eating American requires around 5,400 liters a day; a vegetarian diet in India, on the other hand, requires only 2,600 liters a day. As India becomes richer it is likely that diets there will start to include more meat and thus more water will be required for the food sector alone.

Table 4.1. Estimated Annual Per Capita Freshwater Withdrawal in South Asian Countries and China by Purpose in 2000

Country	Per Capita Withdrawals per Year (m³)	Domestic (percentage)	Industry (percentage)	Agriculture (percentage)
Bangladesh	114	12	2	86
Bhutan	10	36	10	54
China	412	5	18	77
India	497	5	3	92
Nepal	1,189	1	0	99
Pakistan	997	1.5	1.5	97

Source: Gleick (2000: 205). As the water consumption data are provided by states, it is difficult to ascertain their accuracy; some of these are of questionable quality.

Water use in Asia has more than doubled in the last three decades, generally in line with other continents. Even a small change in the populations and economic growth rates of India and China have a disproportionate effect on total freshwater demand in those countries because of their large populations and higher growth rates (World Bank 2002: 204–7). Currently China's consumption of industrial and domestic water is significantly higher than India's (see tables 4.1 and 4.2), and its increasing need to produce more food for its burgeoning population will require intensive cultivation with high-yielding seeds and chemical fertilizer along with new sources of freshwater. India's high economic growth rate is likely to follow the same path as its economic growth rate picks up.

For the purposes of estimating India's future demand, two estimates have been used, one by Messrs Ghosh and Bathija and the other by Messrs Chopra and Goldar (see tables 4.3 and 4.4).

Both projections point broadly to a similar outcome. Messrs Ghosh and Bathija's estimate is based on three scenarios and covers the period to 2050. It shows that the gap between the *low* and *high* scenarios is 15 percent in 2010 and it increases to 48 percent in 2050. Messrs Chopra and Goldar have projected demand for a single period to 2020. Ghosh and Bathija's projection for the year 2025 is closest to Chopra and Goldar's projection under the

Table 4.2. Per Capita Domestic Water Use in 2000 in Mainland South Asian Countries and China

Country	Bangladesh	Bhutan	Nepal	India	Pakistan	China
Estimated domestic consumption [lpcd]	14	10	12	31	55	59

Source: Gleick (2000: 11)

Table 4.3. India's Estimated Freshwater Needs to 2050 by Sector under Alternative Growth Assumptions [billion m³]

Category	2010			2025			2050		
	Low	Medium	High	Low	Medium	High	Low	Medium	High
Irrigation	469	536	556	619	688	734	830	1,108	1,191
Domestic	39.4	41.6	61	47	52	78	59	67	104
Industry	37	37	37	61	67	79	69	81	116
Others	42.1	42.4	43.0	92.0	93.0	94.0	198.0	201.0	305.0
Total	607.5	657.0	697.0	819.0	900.0	985.0	1,156.0	1,357.0	1,716.0

Others include afforestation and ecology. Source: Ghosh and Bathija (2002).

sustainable development assumption. It falls between Ghosh and Bathija's lowest and highest estimates.

They also found that the per capita availability of freshwater per annum has been declining and currently stands at 1,250 CM; it is expected to decline to around 760 CM in 2050. It is affected by evaporation and the evaporation rate is heavily affected by even a slight variation in temperature. The greenhouse effect on evaporation rates is at this stage difficult to ascertain (World Resources 1998: 71), but it cannot be ignored.

The demand for urban water often features greatly in policy considerations, not least because urbanites usually are more articulate and better able to influence policy outcomes in developing countries. Urban population projections and their implications for water demand to 2026 have been made by many researchers.[4] Although there are large differences between the lower and upper ranges of these projections—24.92 BCM versus 40.62 BCM—they unequivocally show that a crisis is looming (Vira and Vira 2004: 301–2).

The biggest source of freshwater is rain and ground water including springwater. Until the third quarter of the twentieth century rainwater was mostly considered pure. However, with the increasing atmospheric pollution the

Table 4.4. India's Estimated Freshwater Needs to 2020 by Sector under Different Scenarios [billion m³]

Category	Business as Usual	High Growth Scenario	Sustainable Scenario
Irrigation	677.30	804.20	768.37
Domestic	67.52	67.52	45.01
Industry	27.91	41.58	27.72
Others	128.19	132.29	125.00
Total	920.92	1,045.59	966.10

Others include power and ecology. Source: Chopra and Goldar (2002).

probability of rainwater getting contaminated by atmospheric pollutant agents (e.g. industrial acid) cannot be discounted. Industrial Europe and some parts of Asia, particularly the People's Republic of China, have already experienced such phenomenon. Purity of groundwater is not guaranteed, as it can get contaminated through naturally occurring chemicals such as *arsenic* or by other harmful minerals or by human activities, particularly industrial pollutants and poor drainage systems.

As in most countries where rainfall is limited to some specific part of the year, rainwater is mostly stored in dams often with multipurpose objectives namely, flood control, using the stored water during non-rainy season and often for generating hydro-electricity. This multipurpose use of stored water is mostly a post–Second World War phenomenon. As we have seen earlier, most Indian dams were constructed in the second half of the last century.

Besides these two sources, recycling of used water and desalinization of seawater are increasingly being pursued to overcome the supply shortage by many countries. Each of these two has specific costs and advantages, as they require large investments to make water usable for the consumers. In the following, advantages and difficulties with each of these sources are considered. Given the climatic uncertainties created by the *greenhouse effect* on the availability of freshwater from the natural sources, and given that wide variety of climate exists in different parts of India, it is arguable whether a single strand of policy will meet India's requirements. It is almost certain that different regions of India require different mixes of freshwater policy.

Rainwater Harvesting

Rainwater collection at the local level has existed in India for a long time. People collected rainwater in the early Buddhist era and the Middle Ages (Appan 2001). The practice is widely followed and encouraged in many countries in Africa, the Mediterranean, and East and Southeast Asia.

Rainwater harvesting is strongly favoured by the environmental movement and NGOs, including those concerned with sustainable development. It is a low-tech solution for water shortages in rural communities and is considered a good source of water, both in developed and developing countries (Waller 1989: 27–36). The main concern in India is whether enough water can be collected through water collectors in four months to meet the entire needs of rural households in particular. As a rule of thumb, 50 LPCD is considered adequate to meet basic domestic needs; an average family of, say, five people would need to collect and store a little more than 61,000 liters of water for use in eight non-rainfall months. This would require large storage tanks and a high-quality collection mechanism. Keeping water from bacterial contamination for such a long period remains a major drawback at this stage. It may be less of a problem in some parts of Tamil Nadu that receive winter rain.

Though huge storage capacity for irrigation and the entire household needs may not be feasible, enough water can be collected and stored to meet domestic needs for drinking, food preparation, and related purposes. If this can be achieved in rural areas, most gastrointestinal and a number of other diseases—major killers, particularly of children—can be prevented.

Besides the low capital cost for a storage tank and collector system, the system is virtually free of running and maintenance costs. This is vitally important since the vast rural majority is unable to pay for water. Another advantage is that individual householders will own the system and can be expected to be careful in maintaining it, knowing that it is their own and that if it is not appropriately maintained no one but themselves will be deprived of clean water (UNEP 2001: 2).

At the micro level another method of harvesting rainwater is the construction of village tanks that can be used mostly for bathing and general washing purposes. In earlier days, tanks played an important role in irrigation in South India, particularly in Tamil Nadu and Andhra Pradesh. Tanks also offer additional income to farmers through aquiculture if they have the necessary skill.

Tanks, however, have a few drawbacks. Siltation reduces their effective storage capacity and there is loss of water through high evaporation and soakage rates. Covering tanks with evaporation protectors may be useful, depending upon individual circumstances and subject to benefit-cost assessments. Another concern is the quality of harvested water. Runoff in the collection ponds will carry many water-polluting agents, both toxic and non-toxic, given that sanitation and garbage disposal mechanisms are non-existent in most Indian villages. A major Indian NGO, however, strongly argues in favour of capturing rainwater at ground level to meet drinking and cooking needs (Agarwal, Narayan, and Khurana 2001: xvi–xvii).

There are other major drawbacks. Tanks are not land-saving structures, and India is a high-density country. More importantly, a little more than 58 percent of the workers in villages in 2001 were either cultivators or agricultural labourers and thus they would not benefit from such community tanks. And it is only realistic to recognize that the medieval social structure in villages fragmented by caste, religion, and feudal power structures may create further tensions.

Until questions of water and land ownership are disentangled, the in-ground tank for storing rainwater is unlikely to be beneficial to most villagers

(Easter and Palarisami 1985: 21–29). Private well owners do not have an interest in tank irrigation, and their desire to make a profit from their wells poses a serious threat to the survival of tanks in many instances. Severe depletion of water tables in Karnataka, Maharastra, and the Deccan Plateau has taken place during the last few decades. Unless groundwater ownership is made clear legally and the administration is irrevocably committed to enforcing the law, vested private interests will find ways of ensuring that they prevail at the expense of community interests (Chandrakanth and Romm 1990: 485–501; Palarisami and Balasubramaniam 2001).

Rainwater harvesting is important for three additional reasons. First, in many parts of India the groundwater is contaminated with toxic chemicals such as arsenic. The level of arsenic found in the groundwater in five districts of West Bengal is extremely high. The origin of the arsenic pollution is geological; it is released into groundwater under naturally occurring aquifer conditions.[5] The possibility that the groundwater might be chemically contaminated never occurred to people, including any of the responsible government agencies, until the rapidly deteriorating situation in Bangladesh attracted worldwide attention.

Following the disastrous consequences of arsenic poisoning in Bangladesh in particular, the World Health Organization (WHO) has lowered the initial allowable limit of 0.05 milligrams arsenic per liter for drinking water to 0.01 mg. Also, in India about 175 places (mostly districts) out of a total of about 593 districts in 2001 were contaminated by minerals (Government of India 2004).[6] While small doses of some minerals for a short time may not be seriously injurious to human health, long-term exposure could be disastrous. Since at the village level the high cost of water processing plants may not be financially feasible, rainwater harvesting is a safe option for overcoming such problems.

Another reason for encouraging rainwater harvesting is its capacity to involve women. Women directly experience its benefits. This encourages them to assert themselves in making decisions. Evidence suggests that in rural Gujarat women are already asserting their views over water policy (Barot and Mehta 2001). This may also help in curbing corruption, which is rampant in the sub-continental countries. As the project is managed by individual or small groups of householders collectively it may not provide opportunities to unscrupulous or corrupt leadership to siphon the funds. An observation by the respected *Economist* is worth noting: "the conviction too often remains that any extra money in the government pot will somehow find its way into the pockets of the country leaders."[7]

The Bangladesh experience is encouraging. Between 1997 and 2001, about 1,000 storage tanks were built with capacities varying from 500 to

3,200 liters. A range of materials was used in their construction; costs varied from $50 to $150 each. Water-quality testing has shown that water can be preserved for four to five months without bacterial contamination (UNEP 2001: 9). Rainwater harvesting, therefore, offers an inexpensive way to solve clean drinking water problems in rural areas, for which governments have often committed huge sums that were invariably found to be inadequate.

Construction of Dams

Rainfall is limited to only three to four months of the year in the sub-continent and this rainfall is not spread uniformly throughout the country. It varies from 150 MM in Rajasthan to 2,000 MM in the northeast. Thirty percent of India receives rainfall between 0 and 750 MM, 42 percent receives 751 to 1,250 MM, 12 percent receives 1,251 to 2,000 MM, and only 8 percent receives 2,001 MM or more (Singh 1979). To ensure availability of water for irrigation purposes, among other purposes, India has constructed a large number of dams during the second half of the twentieth century.

While small dams to store rainwater have existed since time immemorial, the second half of the twentieth century has experienced the construction of mostly multipurpose dams. In the years after independence, India's focus on multipurpose dams to provide water for hydropower and irrigation was welcomed by all. Dams brought immediate benefits along with their costs. Later, Prime Ministers Nehru and Rajib Gandhi questioned the ultimate benefits of dams, but it also remains a fact that Gandhi approved the Sardar Sarovar project against significant national and international opposition (Leslie 2005: 48; Pearce 2006: 35).

The construction of large dams is primarily a post–Second World War phenomenon. Industrialized countries pioneered the modern dam industry. The capacity of most dams to prevent flooding is limited, particularly in Indian conditions. And often it has been necessary to release water from dams because they cannot withstand the high water pressure caused by heavy rainfall and quick runoffs. Thus, instead of preventing floods, dams have sometimes contributed to them. There are many examples.

Floods in the Mahanadi delta between 1960 and 1980 were three times more severe after the Hirakud dam had been built, as the management released water from the dam after incessant rain in the catchment areas. As a consequence, the downstream embankments often breached and people died (Abbasi 1991: 108–9). The 1978 flood in Punjab made 65,000 people homeless because of the forced discharge of water from the Bhakra dam; release of water caused homelessness again in 1989 (Dogra 1992: 38).

Irrespective of whatever preventive measures are taken, some floods will occur. Floods are nature's way of replenishing the productivity of intensively cultivated land, particularly in high-density countries. Floods can reinvigorate soil, as in Bangladesh. In that sense floods are necessary for survival (Paul 1984: 3–19). Strategies should aim to minimize the adverse effects of flooding, particularly where floods are almost annual, as in parts of India and in Bangladesh (Hanchett et al. 1998). The low level of degradation in Bangladesh reflects the positive effects of floods that deliver nutrients to agricultural land. There the cost of flood damage in agriculture was falling, but in non-agriculture it was increasing. Even in bad years the national level of production was not as bad as first thought (Doolette and Magrath 1990: 8, 12; Brammer 1990: 12–22; 1990a: 158–65).

Dams can also run short of water, which needs to be preserved for generating electricity. This often leads to conflict between water-hungry farmers and dam management. For example, in the Telengana region angry farmers threatened to smash electricity-generating machinery upon finding that their valuable crops were dying in the field from lack of water while water was being held back to generate electricity that was of no use to them.[8]

Most multipurpose dams in India, broadly speaking, have been unable to fulfil the rosy expectations painted by the dam-proponents. The failure of most dams to provide enough water for planned irrigated areas, the lack of need for irrigation water during the monsoon months, the poor yield from many canal-irrigated areas, and farmers' inability to recover the additional costs incurred to pay an economic price for the dam water reflect poorly on the water policy singularly based on dam construction (Chambers 1988: 19–20; Jensen, Rangeley, and Dieleman 1986; Mollinga and Bolding 2004: 2).

Many evaluations that justified the construction of dams were conducted using criteria that were often vague, or failed to modify the evaluation criteria to suit country-specific needs (Moreira and Poole: 1993). A few even conveniently ignored what was needed to perform a proper evaluation (Barraclough 2001: 30). Some argued and the World Bank acknowledged that the entire evaluation process has been faulty globally (Doolette and Magrath 1990: 7; Morse and Berger 1992: 41–43).[9] The Indian Public Accounts Committee in 1983 found that largely because of poor feasibility and evaluation studies not a single large irrigation project was completed on time and within budget since independence (Postel 1989: 9; Singh, Kothari, and Amin 1992).

Inadequate cost estimates and cost overruns have been rampant. In 1994 the inflation-adjusted cost overruns of the World Bank's seventy multipurpose dam projects since the 1960s averaged 30 percent, almost three times higher than the average cost overruns on a similar number of other projects.

One year's delay in the completion of a project reduced the benefit-cost ratio by one-third and a two-year delay reduced it by over one-half (McCully 1996: 269–70). The Asian Development Bank has had similar experiences: nine irrigation projects completed by 1980 took an average of 72 percent longer than estimated, incurring a cost overrun of 66 percent (Reppeto 1986: 4). Meanwhile unrealistically low discount rates tilted feasibility reports in support of projects.

There are many examples of poor feasibility and evaluation reports. The Sardar Sarovar Dam claims to have a hydro-capacity of 1,450 MW, but designers failed to mention that the average generation during the initial phase of the project was only 439 MW due to low power production during the long dry season. Eventually the project may become a net consumer of energy instead of a producer (Dharmadhikary 1995). The Bargi dam, the first dam on the Narmada, submerged about 81,000 hectares of farm and forest land and was designed to irrigate an area of about 440,000 hectares. The dam was completed in 1986, but after seven years only about 12,000 hectares—a mere three percent of the planned area—was receiving irrigation water. Its total cost was also many times the budget estimate (Raman 1993; McCully 1996: 87; Roy 1999: 35; 2001: 82).

The environmental appraisal committee of the Ministry of Environment and Forests found that nearly 90 percent of medium and large dams built in India were in violation of environmental and social stipulations required of such projects by the ministry. It found that nearly every irrigation, hydropower, or multipurpose water projects approved in the past 15 years had serious deficiencies in the entire process of evaluating, screening, and clearing projects. More alarming is that, beyond warning some project authorities, the ministry did not take any action to challenge projects that have failed to meet its own conditions (Kothari 1995).

One of the many aspects of dams that have not been adequately considered is unsustainable sedimentation loads. It is estimated that globally in the early 1990s as much as 45 billion tons of sedimentation was building up annually in dams; as a result, cultivation on some 20 million hectares of land has either become uneconomical or impossible (Usher 1997: 3; National Research Council 1993: 337). Even in the United States, the reservoir capacity plummeted from 10.4 acres for dams built before 1930 to 2.1 acres for those built in the 1960s (Devine 1995: 64–74).

A study of India's 11 reservoirs with capacities greater than one CKM found that they were filling with sediments faster than expected, with increases over assumed rates of sedimentation ranging from 46 to a whopping 1,546 percent (see table 4.5). With the climate change phenomenon the sit-

Table 4.5. Siltation in Selected Indian River-dams: Projected and Actual (in acre-feet per annum) and Its Impact on Dams' Life Expectancy

Name of the Dam	Assumed Rate of Siltation	Observed Rate of Siltation	Expected Life as a Percentage of Designed Life
Bhakra	23,000	33,475	68
Maithon	684	5,960	11
Mayurakshi	538	2,080	27
Nizam Sagar	530	8,725	6
Panchet	1,982	9,533	21
Ramganga	1,089	4,366	25
Tungabhadra	9,796	41,058	24
Ukai	7,448	21,758	34

Source: B. Dogra (1986) and L. Brown and E. Wolfe (1984): Soil Erosion—Quiet Crisis in the World Economy, Worldwatch Institute Paper 60, Washington, D.C. Cited in Doolette and Magrath (1990: 7).

uation could deteriorate dramatically in the coming years.[10] One reason could be the failure to take into account the geological changes that have been taking place in the Hindukush-Himalaya region and the soil types in the river basins.[11] An example is the changes in the Brahmaputra course brought about by the 1950 earthquake.

An important reason for the sedimentation build up in dams is soil erosion caused by deforestation in the catchment areas. Even Cherapunji, one of the wettest places on earth, now suffers from regular water shortages caused by the indiscriminate clearing of forests (Rao 1989: 300; US Water News 1995). It is now short of water in non-rainy seasons, as the soil can no longer retain water, which has caused more silt to flow into the Brahmaputra River, contributing to severe and frequent floods in the north eastern plains. There could be other consequences as well: soil in deforested areas in South America frequently desiccates in dry seasons and sometimes becomes saline, compacted, or otherwise unsuitable for agriculture. Dam engineers in India, in all

Table 4.6. Extent of Sloping Land in India by Soil Type and Class (in million hectares)

Soil Type	Rolling to Hills (8 to 30 Percent Slope)	Steeply Dissected to Mountainous (Slopes Greater than 30 Percent)	Total
Luvisols	35.8	2.4	38.26
Acrisols	7.0	7.1	14.14
Nitrosols	12.2	7.6	19.59
Lithosols	1.0	29.0	30.01
Total	55.85	46.25	102.29

Note: due to rounding up, totals may not exactly tally.
Source: FAO/UNESCO Soil Map, World Resource Report (1996) for Land area.

probability, failed to properly assess how the soil erosion occurs in different types of soil and slopes (see table 4.6).

Generally speaking, however, the absence of forests has only a marginal influence on runoff from the Himalayan slopes, as the Himalayan mountain chains are young and erosion rates are usually high: the lower reaches of the Indus River carry up to four times as much silt as the Nile and twice as much as the Missouri (Doolette and Magrath 1990: 2–3; Centre for Science and Environment 1991; Bradnock 1992; Lama 1998).[12]

Many of India's dams have clearly failed to meet their objectives. But in spite of growing questioning of dams around the world, policy makers have remained obsessed with dam construction. That was evident when they jumped with all guns blazing to discredit the work of the International Commission on Large Dams (ICOLD). Yet independent assessment of the Commission's work has nothing but praise for it (Dubash, Dupar, Kothari, and Lissu 2001: 119–28).

The power of the dam industry is immense. Globally, industry organizations have 2,500 members in eighty-one countries. The industry is controlled by forty-one major corporations from industrialized nations. Their power is well known but not always acknowledged. They, of course, endeavour to control the debate (McCully 1996: 248–50). At one industry conference a paper argued that dam opponents had already succeeded in reducing their prestige in the public eye and had created professional difficulties—and went on to argue that the world needed thousands more dams (Usher 1997: 5).[13]

Reasons for such an aggressive stance are not difficult to establish. Most large dam projects receive funding from multilateral and bilateral agencies. Bribes, skim-offs, corrupt and corrupting planning, and the interests of commercial and political actors all shape the funding process (Rich 1994: 190). Indian officials acknowledged that political pressure was applied to pretend that more water flowed down the Narmada than was shown by actual measurement to justify the building of a high dam (McCully 1996: 101–7; Leslie 2005: 45–46).

The role of the bilateral donors is clear. President Richard Nixon acknowledged that the main purpose of providing aid was not to help other nations but to help the donor country (Hancock 1989: 155; McCully 1996: 255 and 272; Usher 1997: 59–60). Private financial institutions also have played a role in this process, mostly in connivance with treasury and central bank officials in developing countries (Truell and Gurwin 1992: 97–98, 165–66, 408, and 422–23; Adams and Frantz 1992: 92).[14]

Failure of multilateral agencies such as the World Bank has also been substantial. For example, the respected *Economist* argued that "too often, project loans are approved only to fail."[15] But the worst of all failures of the dam industry and its governmental proponents are the inadequate assessments of the resettlement costs for projects and the inhumane treatment meted out to the people displaced. The Bank's sociologists pointed out that the *engineering* bias of dam builders was responsible for the lack of time and funds put into resettling people (Cernea 1990: 10).

There is now broad agreement worldwide that the time has come for total rethinking on freshwater management issues, including on the need for large dams (Ingram 1990: v, 1).[16] India has yet to fully grasp this policy shift, insisting as it does on only engineering and construction-oriented freshwater policies. But with demand rapidly outstripping supply, an exclusive focus on supply-side solutions is self-defeating (Ullman 1983: 129–53).

While the Sardar Sarovar Dam proposal could not be stopped, similar proposals in the future will face more intensive opposition from the core of the movement that came into prominence during the save Narmada movement. For it left an indelible mark on the Indian body politic, at least many knowledgeable commentators believe so.

Groundwater Harvesting

Groundwater is the only source of freshwater in many parts of India. In large parts of the country groundwater contains dangerous levels of fluoride, iron, nitrate, arsenic, and salt. In many cases the level of contamination is beyond the permitted level of human consumption. In industrialized countries, particularly in Europe, groundwater is an important source of potable water. There strict application of land-use planning laws prevents surface pollutants from contaminating it. But in most parts of India such laws are non-existent.

India is using its underground water reserves at least twice as fast as they are being replenished (Chakravarty 2004), leading to the drying up of supply (Bandyopadhyay: 1988: 88–95; 1989: 284–92). Unregulated harvesting of groundwater by commercial interests at the expense of the community and the environment, as happened in America in the 1980s, has become a major problem in recent decades (Postel 1989: 19; Plaut 2000: 1, 5; Pearce 2006: 36–37). It is happening in an age when technology makes it possible to regulate groundwater extraction and adopt measures to artificially recharge the aquifers. One researcher has even argued that India is facing a terrible calamity as a consequence of overexploitation of groundwater (Endersbee 2005: 9–10). This is reflected in the water crisis faced by an increasing number of villages, particularly in the Deccan Plateau.[17] Another major concern is the intrusion of saline water in places where an unsustainable level of freshwater is pumped out, particularly in coastal zones. If it remains unchecked, the long-term environmental and economic damage, and damage to health, would be enormous.

Overharvesting is also linked with water markets in India. The lack of an effective monitoring mechanism has encouraged it. It has caused environmental damage, secondary salinization of land, and suffering among those whose water sources dried up. Inadequate recharge also creates drainage problems that lead to waterlogging, soil subsidence, degradation of water quality, and contamination. Some of these problems are especially difficult where the aquifer is effectively a closed system, since these aquifers generally contain a greater concentration of dissolved solids that may not be easy to get rid of (Charbeneau 1982: 957–69; Shah 1993: 88–89 and 206–8; UN 1989: 2–17; Organization for Economic Cooperation and Development 1989: 119–31; Zhaoxin 1990; Winpenny 1994: 55; and Williams 1993).

Contamination also has serious economic consequences, particularly when groundwater-supported activities suffer. If the activities relate to traded goods and services the impact on the regional economy could be serious, particularly in the age of the globalized economy. Because of business's capacity to mount strong political pressures on governments for redress, governments often tend to respond with corrupt policies, as the Arizona case study found (Kelso et al. 1973: 224–57). Hence the importance of knowing the amount of groundwater available and how much can be harvested without damage to humans, animals, plants, or the environment.

Another issue that needs to be considered is water-sharing arrangements with neighbouring countries with land borders, where groundwater-flow management is a more complex issue, as is clear from the experience of Israel and the Palestinians. Many consider groundwater sharing a major stumbling block in the Middle East Peace process, as neither international nor domestic law provides an adequate answer to questions of ownership or rights (Eckstein and Eckstein 2003: 154–61). Political relations between the mainland sub-continental countries, on occasions, are not qualitatively different from those experienced in Middle East.

The UN has been encouraging the judicious exploitation of groundwater in developing countries since 1963 with a long-term perspective that incorporates measures to prevent pollution, contamination, and overuse, among other things (UN 1989a: 2, 12–13, 21). An evaluation of this policy after twenty-five years of operation confirmed its essential correctness. But its success largely depends on the effectiveness of the administration in managing these complex issues. That perhaps remains India's biggest problem in India's freshwater policy.

Desalination of Seawater

The practice of desalinating seawater for human use is very old, with references found in Aristotle's writings. The high costs of desalination and the large energy requirement have limited its application, however. A large proportion of the 8,000 global desalination plants in 1994 were located in countries with a large reservoir of fossil fuel, with Saudi Arabia having 50 percent of the capacity, followed by the United States with 12 percent (Hillel 1994: 251–55). In 1999 India had a limited number of plants with an estimated capacity of 115,509 CM a day (Gleick 1998: table 16).

Desalination technology is improving and production costs have decreased, but the costs are still beyond the financial means of developing economies that need it most (de Villiers 1999: 330, 340; Lora, Sancho, and Sariano 2004: 205). The World Bank estimated in 1993 that the cost of desalination could be as high as $2 CM. Others estimate it in the range of $1 to $1.5 CM. Israel is reported to be able to produce desalinated water for about $0.70 CM.[18] An Indian commentator has claimed that India has developed desalinization technology that can provide freshwater at a fraction of the above costs, but the claim has not yet been confirmed by independent agencies.[19] The costs of processing also vary with the salt content of the water and the methods used. The distribution costs for desalinated water are high except in coastal areas.

Given India's need to import more than two-thirds of its energy needs, desalination cannot be a major policy instrument for augmenting its freshwater supply. The opportunity costs of energy imports in terms of foreign exchange requirements are very high and possibly unsustainable. In 2000, 31 percent of the total value of India's merchandise imports was spent on energy; in 1990 it was 27 percent. Experience shows that countries with existing desalinization facilities were forced to close their plants because of high energy costs. For example, Tuvalu, a South Pacific island country, was forced to close both of its desalinization plants because of high operational costs and high energy requirements (South Pacific Applied Geoscience Commission [SPAGC] 1998: 8).

Although India is a large user of coal (and its use in electricity production increased significantly in the 1990s), the use of coal for desalination purposes is unlikely because of the growing concern for the environment. And India has obligations under the Kyoto Protocol to reduce the emission of greenhouse gases after 2012. Even if it is assumed that desalinization is viable, its application will be limited to coastal areas to minimize cost of water transportation.

Recycling Used Water

Water recycling is gaining acceptance throughout the world, although the pace of acceptance varies. When consumers become aware of the ever-increasing cost of water as a price is charged for it, recycling is likely to receive increasing support. Several issues need to be considered:

- Resistance from the intended users, purely on the grounds of tradition and social taboo, will require community education, persuasion, and inducement to accept it.
- Recycling water requires different degrees of filtration depending upon use-purposes; this will determine the level of investment required for purification to prevent contamination of the food chain.
- Recycled water requires a parallel distribution system, depending upon its use-purposes.

The central consideration in recycling is whether the benefits are on a par with the recycling costs, which include the capital costs. For agriculture the costs are lower, as purification only needs to reach a level that prevents bacteria from entering the food chain. This differs from country to country, as legislative requirements for maintaining health standards vary. Yet those standards are increasingly becoming uniform partly because of the demands of the growing global tourism industry and partly because traders of goods and services are adopting globally accepted health regulations to meet the quarantine and health standards of trading countries.

Barring direct human or animal consumption, the purification requirement for water used in some municipal services such as street-cleaning or watering playing fields, parks, or golf courses is lower than water used in agriculture, for example. The use-purpose will determine the level of purification needed for ultimate users, as the survival times of pathogenic organisms in water and other media vary significantly (SPAGC 1998: 5; Plaut 2000: 7).

The collection of used water in rural areas is unlikely to be economically viable because of the small number of households and the dispersed settlements. The situation is different in urban areas, however, as the cost of collecting water is relatively cheap. In India in 2001, out of a total population of 1.028 billion, only 120.8 million were living in urban areas, about 97 percent of which were living in forty-seven cities and urban agglomerates. Recycling could therefore be considered for forty-seven centers and at individual industrial unit levels, provided that it meets all internationally acceptable standards and is economically and environmentally viable.[20] A study of global megacities from a water perspective found that recycling of water is a necessity to meet their water needs in the future (Niemczynowicz 1996: 198–205).

In India the concept has yet to take off, though. There are a few cases where recycling has been tried; two firms in Chennai established recycling systems that enabled them to double their operating capacity without increasing their demand for water. But Jamshedpur, an industrial town, found that reusing treated industrial affluent was uneconomical (Winpenny 1994: 53).

Given India's shortage of energy resources, the recycling of water has appeal, both financially and environmentally. A study in Orange County in the United States found that a recycling system would use only half the amount of energy required to import the same amount of water. The study further established that even groundwater pumping was more energy-intensive than recycling wastewater (Natural Resources Defense Council [NRDC] 2004).

Managing Irrigation Demand and Improved Water Productivity: Key Elements in a Sustainable National Water Policy

Globally there exists broad agreement that if a catastrophic crisis in the near future is to be avoided, urgent actions must be taken to tackle the *greenhouse effect* issue. Its impacts on loss of habitat, agricultural productivity, human and plant health and general economy are beyond comprehension. At this stage the estimates are tentative and vary, but all broadly agree that time is running out (Dickinson 1989: 5–13; Williams 1989: 83–93; Darwin 2001; Perrings 2003; Hansen 2006; Stern 2006; Connor 2007; Borenstein 2007; and Adam 2007).

For India and particularly for the deltaic parts of the country forecasts are alarming. Experts believe that if the current trend in the greenhouse effect continues 60 million people will be displaced in Kolkata and Bangladesh alone, and that, as the glaciers in the Himalayan region start melting within next five decades, there will be shortage of drinking water in south and north Asia (Gore 2006: 206–7 and 58; Milliman, Broadus, and Gable 1989: 340–45). Another recent study puts the likely number of displaced at a much lower level, although the number itself remains extremely high (Dupont and Pearman 2006: 50).

Population pressures on land in sub-continental countries are enormous. During the last two decades, arable land as a percentage of total land declined marginally in India, but substantially in Bangladesh, and it increased

Table 4.7. Rural Population Growth and Density per km² of Arable Land in South Asian Countries

Country/Region	Rural Population as a Percentage of Total Population		Average Annual Growth of Rural Population 1980–2000	Rural Population Density Per Square Kilometer of Arable Land in 1999
	1980	2000		
Bangladesh	86	76	1.5	1,209
India	77	72	1.6	444
Nepal	94	88	2.0	686
Pakistan	72	63	1.9	403
World	60	53	0.9	524

Source: World Bank (2000); Derived from Table 3.1.

in Nepal and Pakistan (see tables 4.7 and 4.8). Even the increase in Nepal and Pakistan may be due to means such as an unsustainable deforestation rate. This increased pressure will lead to multicropping of existing land, requiring additional irrigation facilities.

Agriculture uses the largest proportion of freshwater globally, and irrigation dominates this use (Shiklomanov and Penkova 2003: 33–37). Getting water from new sources is becoming expensive. The World Bank estimates that the costs of using new sources to procure water could on average be two or three times those of existing investments (Serageldin 1995: 14). An analysis of a sample of World Bank schemes found that cost per-unit of water in new schemes confirms this (Bhatia and Falkenmark 1991, cited in Winpenny 1994: 4). As about 90 percent of India's freshwater demand is for the agricultural sector, that sector could not bear this increased cost and still remain competitive, given low farm productivity (Goyal 2002). The current water use–efficiency rate in India is somewhere between 25 and 40 percent only (Rosegrant 1997: 4).

Land under irrigation increased by 93 percent worldwide during 1961–1997. In Asia it increased by about 108 percent and in India by 131 percent during the same period (see table 4.9). The UN projects a 50 to 100 percent increase in use of irrigation water by 2025 (World Meteorological Organization [WMO] 1997: 1, 21). The speed at which land under irrigation increased globally in the early post-war years slowed significantly in the 1970s and 1980s (Ghassemi, Jakeman, and Nix 1995: 12–13). Most likely the main reason for this slow growth is the increased salinity and waterlogging caused by canal-irrigation projects.

In the 1980s worldwide, about 150 MH, nearly two-thirds of the total irrigated area, required some form of upgrading to remain in good working order

Table 4.8. Changes in the Percentages of Arable and Irrigated Land in South Asian Countries

Country	Land Area (in 1,000 Square KM in 1999)	Arable Land as a Percentage of Land Area			Irrigated Land as a Percentage of Cropland		
		1980	1999	Changes 1980–99	1979–81	1990–91	Changes 1979–81 & 1999–01
Bangladesh	130	68.3	62.2	–6.1	17.1	46.1	+29.0
India	2,973	54.8	54.4	–0.4	22.8	33.6	+10.8
Nepal	143	16.0	20.3	+4.3	22.5	38.2	+15.7
Pakistan	771	25.9	27.5	+1.6	72.7	81.7	+9.0
World	130,100	10.2	10.5	+0.3	17.7	19.9	+2.2

Source: World Bank (2002).

Table 4.9. Growth in Irrigated Land in South Asia and China during 1961–97 (1,000 hectares)

Country	World	Asia	Bangladesh	Bhutan	India	Nepal	Pakistan	China
1961	138,813	90,166	426	8	24,685	70	10,751	30,402
1997	267,727	187,194	3,639	40	57,000	1,135	17,580	51,819
Changes 1961–97	93%	108%	754%	400%	131%	1,521%	64%	70%

Source: Food and Agriculture Organization 1999: www.fao.org; cited in Gleick (2000).

(Rangley 1986: 355–68), because of increased salinity in canal-irrigated land, caused by a lack of proper drainage systems. Farmers usually ignore the construction of a good drainage system while accessing canal irrigation, as a good drainage system costs large sums of money (Goldman 1994, cited in McCully 1996: 170). In the long run it causes land to loss its productivity when it is too late for any correction. In India about one-half of the total land area was affected by soil degradation in 1990; the figures for Bangladesh and Pakistan were 7 and 17 percent (Economic and Social Council for Asia and Pacific [ESCAP] 1992: 111). The Pakistan estimate is substantially higher than earlier estimates (Jensen, Rangeley, and Dieleman 1986).

The Indian land degradation started when the British introduced large-scale canal irrigation at the beginning of the nineteenth century, although a British government chemist warned that it would create enormous damage to thousands of hectares of crops and that the land would look like a salt-covered desert (Whitcombe 1972: 72). Only six years after the Ganges canal system was opened by the British in 1854, the government had already found that it was of little use to the local population (Pearce 1992: 67–71). In large parts of Punjab, Haryana, and elsewhere, a short-term view of production has led to excessive use of fertilizer and salination of cropland. Experience from Pakistan and parts of northern India points to an irreversible salinization of land because of increasing canal irrigation. Efforts to reclaim soil are being made, but the damage cannot be reversed (Burke 2002).

Canal irrigation is also land-hungry. In India an area equivalent to between 5 and 13 percent of newly irrigated land is typically lost to reservoirs, canals, and drainage infrastructure (McCully 1996: 166–67). Seepage through canal line-ups during water transportation is also a major problem (Winpenny 1994: 71), as are huge evaporation rates. Canal seepage can be partially minimized, but that calls for increased investment in construction and continuous maintenance of the canal shorelines. India's record on maintenance of public utilities is extremely poor. For example, even a showpiece project like the Farakka Barrage lost one of its lock gates within three decades of its construction due to poor maintenance creating a huge crisis in sharing water with Bangladesh.[21]

Compared to canal irrigation, evidence shows that productivity in *well-irrigated* lands is on average nearly double that of canal-irrigated projects. This could be due to the controlled use of well water and its lower level of salinity, as canal water carries with it more salt collected from the washout of other land. Nevertheless, Indian policy remained focused on investment in canal irrigation, which required the construction of dams that not only failed to achieve their basic objectives but have left behind a legacy of unsurpassed cultural destruction, disease, and environmental damage (Goldsmith and Hildyard 1984: 27–48; vol. I; Barber and Ryder 1993: xxi).

Independence in India did not bring any noticeable change in the policy of canal irrigation. Engineers, planners, and administrators followed virtually the same route. Their neo-colonial education, class arrogance, and limited appreciation of the water-management knowledge passed on by the generations of hands-on practitioners did not help the cause of agriculture either (Sengupta 1985: 1919–38). By following the same old policies and ignoring warnings by experts, the longer-term interest of the country was sacrificed (Bhumbla 1984).

A little more than eight BCM of water was used to irrigate one MH of land; on average, 60 to 75 percent of water is lost to evaporation or runoff. A 10 percent increase in the use-efficiency rate could add 14 MH of irrigated land (see table 2.2, page 33). If the efficiency rate is brought to the average industrialized-country level, a savings of 50 to 250 BCM is achievable. Pakistan, a water-scarce country, could also achieve substantial savings by bringing about two MH of additional land under irrigation (Postel 1993a: 60; Saleth 1996: 234). But this productivity improvement cannot be made without increased investment in technology.

New irrigation methods that allow water to directly and intermittently feed the targeted plants and bypass the atmospheric exposure that creates wastage and low water productivity provide opportunities to increase water productivity enormously, before there is any need to make costly investments to make additional water available (Repetto 1986: 21–32; Hillel 1987: 44; Postel 1989: 36–43; Rosegrant 1997: 4; Gleick 2003: 187–98). These savings could be achieved firstly, by integrating land-use with the value of the cultivated crops; and secondly, by introducing new irrigation techniques. Of course, large investments will be required to introduce new technology, but these will contribute to preserving valuable natural resources and will be environmentally more friendly.

Linking agricultural land use with the economic value of cultivated crops can contribute to improving water productivity. For example, a large quantity of water can be saved per comparable unit of production by replacing rice with potatoes (see table 4.10). Current agricultural practices in India ignore this totally, however, as is clear from sugar cane cultivation in Maharastra.[22]

Drip irrigation technology that came into existence in the 1970s has failed to improve water productivity significantly since it allows the drift of droplets before they reach the intended plants and thus contributes to higher rates of evaporation (see table 4.11). Consequently since the mid-1970s only one

Table 4.10. Amount of Water Required (in m³) to Produce One Kilogram of Selected Food Items

Crop	Potato	Wheat	Alfalfa	Corn/Maize	Rice
Water	500 to 1,500	900 to 2,000	900 to 2,000	1,000 to 1,800	1,900 to 5,000

Source: Gleick (2000: 78). Compiled from a number of crop-specific studies.

percent of the world's irrigated area has been brought under this system (Postel 1993: 60).[23]

In recent decades a new mechanism—generally known as low-energy precision application sprinklers (LEPAS)—has reduced demand for irrigation water substantially (Hillel 1987: 31). It takes into consideration factors such as climate, soil, and type of crop, and is alleged to have raised water productivity very significantly (Postel 1999: 186-7).

Davies and Dry's (2003) research is showing promising results for a dramatic water-saving mechanism for non-cereal crops. In an open and competitive economy, the agricultural sector loses out, because in value-added terms industrial, domestic, and recreational uses create more value per CM of water (Shiklomanov 1993; Elmusa 1997: 211). Hence the critical need to relate the real cost of water to the value of the product as far as practicable.

In conclusion it should be emphasized that freshwater management policy should adopt a holistic approach that includes improving water productivity and harvesting water at the lowest possible costs, both financial and environmental. Rainwater collection, better groundwater management, desalinization, and recycling are the key areas that need to be brought within the purview of water policy. A few irrigation engineers, including some in India, argued long ago that this approach to irrigation was essential, and that it should include a whole range of operations from the drawing board to crop sale (Zimmerman 1966: 1-31 and 87-91; Chopra 1986: 97-118).

Table 4.11. Productivity Gained by Shifting to Drip Irrigation from Conventional Irrigation for Selected Agricultural Items in India in the Mid-1990s

Changes (Percentage)	Banana	Cabbage	Cotton	Sugarcane	Sweet Potato	Tomato
Crop yield	52	2	25 to 27	6 to 33	39	5 to 50
Water use	−45	−60	−53 to −60	−30 to −65	−60	−27 to −39
Overall productivity	173	150	169 to 255	70 to 205	243	49 to 145

Source: Gleick (2000: 84).

None of this will be easily achieved, particularly in India, not only because many people do not yet consider water a scarce economic resource but because new measures require significant capital investment. Differences in cost between measures can be high. Large-scale capital investment is normally beyond the reach of most small farmers and their cash-starved state governments. But without improved productivity, better management, and an increased price for agricultural produce, investment in the agricultural sector becomes unsustainable.

Notes

1. Samuel Taylor Coleridge (1772–1834), English author, poet, and critic wrote "The Rime of the Ancient Mariner."

2. See Falkenmark and Lindh (1975: 1–2 and 21–29); Falkenmark et al. (1998: 148–54).

3. It is recognized that a plausible estimate of the water demand coefficient requires the overcoming of intractable informational and technical difficulties. As developing economies mostly will not have the required data, estimates should be considered as approximations only.

4. Agencies and people are: Tata Energy Research Institute; Ministry of Urban Development, GoI; NCIWRD; Chopra and Goldar; and Bhaskar and Shiraz Vira.

5. See http://bicn.com/acic/ (November 29, 2004).

6. See www.censusindia.net/t_00_001.html (June 28, 2004). The number of affected districts should be taken as indicative only. GoI data show "place of occurrences" and identified those by district names only. The number of affected districts is based on this assumption.

7. *Economist* (1993), "Borrowed Time: African debt"; May 22; p. 44.

8. *Economist* (2002), August 24; p. 49.

9. International River Network President's address at the 1991 Vienna Conference (Pearce 1992: 140). It is impossible to discuss the magnitude of collective failure of all parties, particularly the GoI and the relevant state governments in a few lines or paragraphs. Also, see Professor T. Scudder's Report to the World Bank *The Relocation Component in Connection with the Sarda Sarovar (Narmada) Project* (1993) in conjunction with M. Morse and T. Berger (1992: 41–59).

10. It is estimated that by 1986 around 1,100 CKM of sediment had accumulated in the world's reservoirs, consuming almost one fifth of the global storage capacity. Global climate change is expected to create more serious consequential effects, as the amount of sediment carried into a reservoir is at its highest during floods. Given that global warming is expected to cause more intense storms, there is a significant possibility of such situations arising (Mahmood 1987: 8–9 and 32–37; Williams 1989: 83–93).

11. The region extends from Pakistan across northern India, Nepal, Bhutan, and China. The range is in continuous state of formation as tectonic drift drives the In-

dian Plate under the Eastern Plate at the rate of 5 CM per year, lifting the Himalayas 1 CM per year in altitude (Chapman 1992: 10).

12. Other interesting works on this issue are: J. D. Ives and B. Messerli (1989), *The Himalayan Dilemma—Reconciling Development and Conservation*, the UN University, Tokyo; P. Tapponier, G. Peltzer and R. Armijo (1986), "On the mechanics of the collision between India and Asia" in M. P. Coward and A. C. Ries (eds.) *Collision Tectonics*; Oxford, Blackwell Scientific Publications.

13. Pircher at the 1992 British Dam Society Conference at Sterling University, U.K.

14. The collapsed Bank of Commerce and Credit International is an example. P. Truell and L. Gurwin's book exposes the unholy alliance between various parties in this entire episode.

15. *Economist* (1993), "H—Street Blues," May 1; pp. 83–84.

16. Democrat Congressman Udall in his foreword to Ingram's book (1990: 1) acknowledged that as a core activist for the Central Arizona Project he attracted the enmity of environmental groups, but later built a career as an environmentally sensitive legislator, winning accolades for shepherding the Alaska wilderness and park legislation through the Congress. In 1987 he expressed doubts publicly as to whether he would be in favor of the Central Arizona Project were he to relive that part of his career.

17. In 2003 the Kerala High Court ordered Coca-Cola to stop pumping out the local water and find an alternative source of water for its bottling plant (Pearce 2006: 44).

18. Information communicated to Hillel by a former Water Commissioner of Israel. As for other countries, Abu Dhabi produces 40,000 CM/day at a cost of $0.70 / CM; Larnaca in Cyprus with similar capacity produces at a cost of $0.73 / CM; and Tampa in the United States with a capacity of 100,000 CM produces at a cost of $0.46 / CM (Lora et al. 2004).

19. *Statesman Weekly* (2004), "Water Security: Introduce Desalinization Technology in Dry Coastal Belts"; July 24.

20. The estimated total includes the projected population of the Indian state of Jammu and Kashmir, where an actual count could not be undertaken due to security concerns. Class I towns [called cities] are those with a population of 100,000 and above; Class II: 50,000 to 99,999; Class III: 20,000 to 49,999; Class IV: 10,000 to 19,999; Class V: 5,000 to 9,999; and Class VI: less than 5,000.

21. *Statesman* (2007), "Farakka lock gate washed away"; February 21.

22. It takes up to 2,000 CM water to produce the sugarcane from which one ton of sugar is extracted. This is based on an extraction ratio of 10 to 12 percent (Elmusa 1997: 194).

23. Although no precise data on the location of drip-irrigation areas are available, it is widely accepted that most are located in developed economies.

CHAPTER FIVE

The River-Linking Project

India is not blessed with equitable distribution of water resources. It has places with about 150 days of rain a year, and deserts where it rains only five days a year. Except in small parts of southern India, 80 percent of the rainfall is concentrated during the period June to September. Generally speaking, rainfall and weather statistics alone fail to provide a clear picture, however, as the spatial distribution and timing of rainfall are equally important for crops. Even in an official drought,[1] crops can still survive if enough rain falls at the critical stage of the plant's growth cycle.

Sir Arthur Cotton, a well-known British engineer in the 1850s, proposed to link a number of Indian rivers to bring more land under irrigation. In his initial years in India he had praised India's existing irrigation system, but with increasing pressures to earn extra revenue he changed his position, despite warnings by British soil experts of the possibility of soil degradation caused by excessive irrigation. Those warnings have been overlooked.

To overcome water shortages in the southern parts of the country, irrigation engineer Sir Visveswaraya first raised the idea of river linking in pre-independence days. After independence, K. L. Rao, an engineer-turned-politician, and Captain Dastur, an aircraft pilot, raised the idea of selectively linking northern (Himalayan) and southern (Peninsular) rivers. The CWC and other experts did not find the idea worthy of consideration (Bandyopadhyay and Perveen 2004: 5307–16), but it was revived in the 1982 and formally incorporated in the 2002 national water policy (Singh and Shrivastava 2006: xiii). The government then appointed a task force in 2003 to implement the project.

The objective of the river-linking project is to store the surplus water from the Himalayan Rivers during the rainy season and transfer this water to the Peninsular Rivers. Flow volumes in these two major river systems in India vary significantly. The snow-fed Himalayan Rivers flow (though variably) throughout the year; the Peninsular Rivers are primarily rain-fed, and their flow levels fluctuate considerably between the dry and wet seasons.

The project proposes to construct 200 large dams to store water, and to connect ten major rivers and several smaller rivers through a total of thirty links. It is expected to deliver 173 BCM of water, irrigate 35 MH of land, and eventually generate an additional 34,000 KW of hydropower. About 450,000 people will be involuntarily displaced by the project, estimates show, and about 79,300 hectares of forest land will be submerged. Officially, the project is estimated to cost about $125 billion (at current prices).

The task force indicated that the project would be completed within 12 years, although none of the pre-feasibility or feasibility reports are available for public scrutiny. In 2005, some 14 feasibility reports were put on the National Water Development Authority (NWDA) website along with the comments of state governments.[2] These reports cover some minor canal-link projects only, with an investment requirement of about $19 billion. An independent study, however, claims that pre-feasibility reports of all proposed links have been completed, along with feasibility reports for eight links (Sowani 2006). The government's approach is baffling, to put it politely. A recent press report indicates that an MoU for preparation of a detailed project report of the Ken-Betwa link was signed trilaterally by the Centre and governments of MP and UP in 2005,[3] although none of the basic pre-feasibility reports have been examined or deliberated openly yet. Large sums are spent to complete details of small sub-project components, whereas the critical issues have been kept away from the scrutiny of the stakeholders and public at large. It is also strange that the authorities have published feasibility reports of some minor components before making reports of key components available.

Irrespective of the merits of the comments of state governments on the reports, such technical matters should be considered by all stakeholders openly and transparently, not by a few bureaucrats or politicians or their chosen technocrats only. Independent researchers, experts, and commentators have insisted that projects of such nature firstly, should be assessed professionally right from the stage of pre-feasibility studies to establish the facts beyond any technical doubt; secondly, these assessments should be undertaken transparently; and thirdly, democratic participation of stakeholders in the planning process should be ensured (Prasad 2006; Padmanabhan and Poongavanam 2006).

Stakeholder involvement is a must, as time and again in the past governments of all persuasions and vested interests have deliberately manipulated feasibility and benefit-cost reports of projects. Uliveppa, a university researcher lamented that whenever approached the Government Task Force took the position that technical reports would not be of interest to the public (Uliveppa 2006). Given that the GoI's record with feasibility and environmental impact studies has been dismal in the past, policy adopted by the Task Force certainly raises many doubts. The Indian print-media is also vocal that the feasibility reports must be considered in their totality and be evaluated and debated publicly to avoid past policy mistakes.[4]

Like all policies, water policy is a political statement of the government that involves consideration of complex social, political, economic, and environmental issues. The proposal is part and parcel of the government's water policy. More specifically a number of critical issues concerning the proposal need to be examined thoroughly before taking a final decision, and a failure to do so objectively may lead to catastrophic consequences (Chinnammai 2006). These issues are: environmental, social, economic, financial, and political (e.g., inter-state and international disputes).

For India both strands (national and international issues) of the political issues are complex, if experience of past six decades is any guide. During the last six decades the government with all its might has not been able to resolve a single major internal water conflict satisfactorily yet. Even the latest Cauvery Tribunal decision has raised the political temperature, and it is not clear yet whether the central government will be able to solve the issue at the political level. The outcome is anybody's guess, because the demands of coalition politics make most political decisions simply unpredictable.

In India there are a few states that are certain to be adversely affected, as the proposal would directly affect control of the acquisition and distribution of water. This and the negative consequences of the project in the international arena need to be addressed to the riparian countries' satisfaction (Salman and Chazournes 1998: 167; Chandler 2004: 207).

While these are critical elements, the basic parameters of water-policy are set by the climate, the drainage basin, and the ecosystems. None of these three elements may be artificially tampered with unless one is absolutely certain about the consequences of the new initiatives taken to meet other policy objectives. By proposing to proceed with the project before all the necessary information had been collected, analyzed, and debated publicly, the GoI ignored all these. Furthermore, the importance of stakeholder participation in policy development has been totally ignored, to say nothing of ecological and environmental concerns. In the process GoI is also ignoring the basic pillars of contemporary water management policies, namely efficiency, equity, and sustainability (Elmusa 1997:175).

One Policy Does Not Fit All

India is an extremely diverse country physiographically and climatically. And yet the project proposes to deal with water problems using a broad-brush approach. Physiographically India is divided into seven regions and each region has its unique physiographic and climatic characteristics:

- The Northern Mountains: comprising the Himalayan ranges;
- The Great Plains: mostly cutting through the Ganges-Brahmaputra river system. About one-third of the land covered by this river system is in the arid zone of western Rajasthan; the remaining two-thirds are mostly fertile area;
- The Central Highlands: the area between the Great Plains and the Deccan Plateau falls in this region. The area mostly consists of the wide belt of hills in the Aravalli ranges in the west and terminates in the steep escarpment in the east;
- The Peninsular Plateaus: covers the Western and Eastern Ghats, North and South Deccan Plateau, and the Eastern Plateau;
- The East Coast: the land area of about 100 to 130 km wide between the Bay of Bengal and the east part of the Eastern Ghat mountain;
- The West Coast: the land area of about 10 to 25 km wide between the Arabian Sea and the west part of the Western Ghat mountains; and
- The Indian Islands: located in the Arabian Sea and Bay of Bengal.

The Himalayan basin has three sub-systems: the western, central, and eastern. The Indus group of rivers is in the western system, in which India and Pakistan have primary interests. The central system includes the rivers of the Ganges system, in which India, Nepal, and Bangladesh have primary interests. The eastern system includes the rivers of the Brahmaputra system; China, Bhutan, India, and Bangladesh have primary interests there. The proposal therefore affects the ecological, environmental, climatic, and economic conditions of these sovereign countries. The main Peninsular Rivers namely, Narmada, Tapti, Sabarmati, Damodar, Mahanadi, Godavari, Krishna, and Kaveri originate in and flow through Indian territory (Misra 1970: 139–40).

Of the thirty river links through which the project would redistribute water, fourteen would be constructed in the Himalayan part and sixteen in the Peninsular part. The former would involve dealing with international rivers.

The project envisages the construction of storage dams in both India and Nepal, mainly on the principal tributaries of the Ganges and the Brahmaputra. Only about 40 percent of the usable waters of the Ganges-Brahmaputra-Meghna system are currently used, compared to 50 to 90 percent in other river basins. The proposal to build 200 new dams needs to be seen in the context of the strong movement globally towards river restoration. For example, nearly 500 dams in the United States alone were removed by 2002 with the objective of restoring the dammed rivers (Gleick 2000: 277–86).

The proposed dams will change the entire geomorphology of India, as they will interfere with the natural flows of most Indian rivers. They will also disturb the natural equilibrium between land and water systems that has existed for thousand of years.

Given all that, it is essential that those aspects of the UN's declared eleven challenges to the management of the dwindling freshwater supply globally that relate to nature be examined as thoroughly as possible (UN-WWDR 2003: chapters III and IV).[5] Three of these eleven challenges bear on natural phenomena:

- Will it adversely affect "ecosystems for people and planet"? (pp. 130–34). This is critical, as "a wide range of human uses and transformations of freshwater or terrestrial environments have the potential to alter, sometimes irreversibly, the integrity of freshwater ecosystems";
- Will it contribute to "mitigating risk and . . . coping with uncertainty?" (pp. 272–77); and
- Will it contribute to "governing water wisely for sustainable development?" (pp. 370–83). Freshwater is a finite natural resource, and the issue has virtually taken center stage in this debate.

The first issue is an extremely critical issue. The proposal will severely impact on the freshwater and natural ecosystems of almost entire India. Anecdotal evidence indicates that everywhere there have been changes in climate-behavioural patterns, and mostly these are unpredictable. The impact of such changes cannot be measured effectively, but the signs are ominous. In the Himalayan Region glaciers are receding and globally everywhere snowfields are not gathering as mush snow as in earlier times.

More generally, international studies on the effects of climate change on the environment and society need careful consideration, as these might foretell what could happen in this region as well. Scientists are now confidently predicting that droughts in the twenty-first century are likely to be more intense and qualitatively different from previous droughts (Schindler 2003:

164–68). There are two major reports: one by the UN and the other by a former World Bank economist (Intergovernmental Panel on Climate Change—IPCC, and the Stern Review on the Economics of Climate Change). Besides these two highly applauded reports, let's consider the findings of another important study in Canada. In Alberta during the past 2,000 years, on average there were three to four major droughts in each century, and at least one lasted for more than ten years. Temperatures in that region increased by 10 to 40° C over the past fifty to one hundred years, resulting in increased evaporation and seepages from the land. The recent four-year droughts were more intense than any in the last 120 years.

As far as the challenge of mitigating risk and coping with uncertainty goes, India could have done better. For example, while only one of the fifteen major droughts in the last forty years was in India, the Indian drought (1965–1967) recorded the largest number of deaths. That indicates major shortcomings in drought management practices (UNWWDR 2003: 226–27, 272, and 275). Industrialized countries have policies in place that enable them to minimize the risks from such disasters and consequently they can cope better with them.

As for sustainable development, even with the existing irrigation schemes (irrespective of their sizes), productivity has remained much below the projected rate. That indicates that water is only one of many components that are required for increased productivity. Other factors are land reform, land salinity, prices of agricultural commodities, credit, land-management practices, and infrastructure including marketing facilities. These issues need to be addressed objectively.

Opposition to the Project

The project will interfere with the natural flows of rivers in an unstable geological environment where a large number of storage dams will be constructed. It will also affect the ecological balance and physical forms that have evolved over thousands of years. The effects may not be felt or detected immediately, but the cost to future generations could be enormous—hence the need to take time and understand the environmental effects as thoroughly as possible. Well-known experts such as Williams (1993) and Razvan (1991) warned about the unknown dangers in massively disturbing the natural equilibrium of ecosystems established over millions of years. Experts such as Pearce supported this position (1992: 141, 347–48).[6] In undisturbed natural systems, water usually performs a number of functions, in the process accumulating as soil moisture and in wetlands, lakes, and rivers. It moves in the underground aquifers, evaporates, and affects the temperature (Wescoat, Jr., and White 2003: 5). These activities have a profound influence on the riverine systems of any country. But there is little evidence that the GoI has given any consideration to these issues.

The project is opposed by many, both within and outside the country. Bangladesh at all levels and Nepal at NGO levels have already launched strong campaigns against it.[7] Bangladesh shares 60 percent of the Gangetic delta with India, and experts on the Indian side have predicted large-scale physical instability in the deltaic region, with disastrous human and economic consequences. An Indian interdisciplinary group strongly argued that, technically, the project might be doable but not desirable, as the proposal is beset by several potentially dangerous consequences, many of which were unforeseeable (West Bengal Academy of Science and Technology 2003). Also in 2003, fifty-eight experts published an open letter to the Prime Minister, President, and the Chairman of the Task Force highlighting the dangers of this project and to consider all consequences of it thoroughly and transparently. In 2006 another group of academics and researchers published a volume with similar views.[8]

Among many experts, Bharat Singh, a noted engineer and academic, argued that the project was based on knowledge of European rivers rather than the realities of rivers in India (Bandyopadhyay and Perveen 2004: 5307–16). Even in Europe countries are now shying away from such grandiose projects; Spain's retreat from the Ebro river project is an example.[9]

At the institutional level the GoI's cavalier approach to the project is also questioned. The International Water Management Institute (IWMI) is considering it seriously. It has announced a three-year study to determine all aspects of its feasibility and to generate a national debate on the inter-basin transfer of water. It argues that it is the largest infrastructure project in the world and that the study will ascertain its cost-effectiveness and sustainability

Experts from twelve countries at the Brazil climate-change conference in 2004 also heard arguments that Indian states such as Bihar, West Bengal, Meghalay, and Assam would be adversely affected by this project in addition to Bangladesh. While the opposition at the conference was led by Bangladesh, it would be counterproductive to dismiss it on political grounds. Such distrust is the single most important reason why both feasibility reports and EISs need to be conducted transparently and disseminated as widely as possible.[10]

Notwithstanding the above, the project is described, by the President down to local leaders in regions with lower rainfall—without any substantiation—as the perfect win-win solution that will solve the twin problems of water scarcity in the western and southern parts of the country and floods in the eastern and northeastern regions. It is not, therefore, surprising that the lower riparian country Bangladesh is very apprehensive of the project.

Where Is the EIS?

Given the complex nature of the river-linking project, it is essential that feasibility and environmental assessment reports are undertaken on the basis of both known and unknown factors. The process must also be transparent. Transparency would demonstrate to supporters and opponents alike that India is not hiding anything and that all the consequences have been taken into account. This would help limit disagreements. It needs to be kept in mind that in the past Indian officials and the leadership in the sub-continent were often accused of creating a fear-psychosis in the minds of people, including officials and politicians in the small adjoining countries (Upreti 1993: 188–89).

Producing an environmental impact statement (EIS) based on existing knowledge is relatively easy. The more difficult part of the assessment will involve those aspects that cannot be undertaken with existing knowledge or methodology but nevertheless are essential to minimize the risk of making decisions without any inkling of what could go wrong—particularly when climate change and its consequences remain hotly debated.

It is also important that EIS and feasibility reports are not cloaked in technical jargon, since that limits wider community and stakeholder participation in policy deliberations. Global experience shows that vested interests often do not hesitate to use technical jargon to help their viewpoints prevail in the public domain. Such practices allow the unscrupulous to hide many important matters from public scrutiny. There have been plenty of examples in India during the last six decades.

Most pre-1994 water related projects in India were carried out without any detailed and transparent EIS. The situation has not changed much over the years and stake-holder participation is not yet universally assured. Furthermore, even in cases where conditional EIS approvals were given for projects, project authorities in 90 percent of cases had not complied with the conditions that were attached to the projects (Kothari 1999). The situation has not changed at all, if the media reports are to be believed.[11]

No matter how good the quality of an EIS, it is now generally accepted that EISs can never be a perfect exercise because what is presented by one set of experts is often a social construct that can be deconstructed and reconstructed by other experts. Flyvbjerg, Bruzelius, and Rothengatter (2003: 49) identified a number of reasons for the poor accuracy rate of EISs by using findings of the other studies. However, it is broadly accepted that there are generally three major deficiencies in EISs:

- A lack of accuracy in impact predictions;
- The narrow scope of impacts and their time horizons; and
- Inadequate organization, scheduling, and institutional integration of the environmental-impact assessment process in the overall decision-making process.

The failure of many EISs to make reasonably reliable assessments has led to a plethora of studies to identify what options are available to improve EISs. But most agree that by undertaking post-audits of EISs, defects could be minimized.

The most pertinent task for an EIS, therefore, may be not to predict with complete accuracy but to define appropriate goals and then set up the organization that can effectively adapt and audit the project to achieve them, if it is found otherwise acceptable. The main obstacles are the absence of mandatory, institutionalized requirements for post-auditing and indifference among authorities and developers to such audits (Flyvbjerg, Bruzelius, and Rothengatter 2003: 57–64). Initially, at least, the EIS must be undertaken transparently and under alternative assumptions, which will enable all stakeholders to understand the consequences, if any of the assumptions is found to be not to the mark.

A generalized version of what an EIS should usually contain for projects is included in Appendix Two. The reader should particularly remember that issues concerning the global climate change due to the greenhouse effect is not included in the appendix, firstly, because the subject is still considered fluid by scientists and many policy makers, and secondly, a large number of literature separately exists which deal with this issue effectively that cannot be undertaken in the context of this study.

The River-linking Proposal and the Gangetic Delta

This issue needs specific examination because the project will have disastrous consequences (ecological, environmental, and economic) for the deltaic Bengal, both for India and Bangladesh, Bihar, Orissa, and possibly the entire North East region, coastal Andhra, and maybe even for Nepal and Bhutan.

Geologists and soil scientists believe that the Gangetic Delta, the largest in the world, is not a stable environment. Some argue that normal deltaic activities have been complicated by recent geological events, such as the 1950 earthquake, which increased sediment loads in the rivers in the region, particularly the Ganges, Brahmaputra, and Meghna. These events along with substantial deforestation and bad agricultural practices have had severe effects on the ecological balance in the entire deltaic area, including changes in the morphology of the channels and increased sediment flows and salinity (Crow 1995: 129–36).[12]

The effects of such changes in the Gangetic Delta could be enormous in eastern India and Bangladesh (Milliman, Broadus, and Gable 1989: 340–45).[13] The 20,000-square kilometer (SKM) delta stretches across these two nations; 60 percent of it is in Bangladesh, the rest is in India. The sea levels in the deltaic area have been rising at an average rate of 3.14 centimeters a year over the past two decades, much higher than the global average of two millimeters a year. Within the next fifteen to twenty years another 15 percent of the habitat area could be lost, displacing more than 30,000 people; most of them will possibly join the impoverished pool of landless labourers (Hillier 1988: 77–79). A BBC documentary team visited an Indian island (Ghoramara) in the deltaic area and found that some parts of the island had already gone under the water and some parts were already lost to erosion. Long-term inhabitants are worried that in not too distant a future they do not know where they will go, as they are gradually losing everything they had.[14]

The Ganges-Brahmaputra-Meghna basin, a network of 230 rivers and streams, carries sediment loads of 1.7 to 2.4 billion tons on average, more than any other river system in the world (UN 1995: 20–28). Robbing rivers of sediments by establishing a large number of dams upstream adversely affects the stability of long stretches of coastline facing erosion from waves (McCully 1996: 36–41). The deltas also act as a major buffer against nature's fury. For example, the 2004 tsunami showed that mangroves can provide a more effective defense against tidal waves than man-made walls (Abraham 2005).[15] While southern parts of Bangladesh and parts of West Bengal, Orissa, and coastal Andhra in India have not been hit with the force of the 2004 tsunami, it is believed that the deltaic mangroves have, at least in part, mitigated cyclones and tidal waves in these areas. The mangrove forests in the delta also provide a natural breeding ground for a large variety of fishes that provide livelihood to an enormous number of families in the entire region.

The effects of dams on erosion in the downstream areas of the coastal belt also need to be assessed. The precise reasons for the continuing erosion in three Indian districts—Malda, Murshidabad, and Nadia—following the completion of the Farakka barrage have not yet been definitely established. These erosions are creating panic among the poverty-stricken people, and much of coastal south Bengal could become uninhabitable as persistent typhoons and rising sea levels cause further coastal erosion, contaminating inland freshwater and creating a drinking water and humanitarian crisis (Schwartz and Randall 2003). An Australian study shows that there is a strong causal relationship between dams and coastal erosion.[16]

Overuse of fertilizer, discharge of untreated sewage, and ever-rising emissions from vehicles and factories are producing alarming and sometimes irreversible effects on rivers and the coastal environment all over the world. These effects could lead to a rapid expansion of the global dead zone. Dead zones, the marine equivalent of the ozone hole, have increased enormously throughout the world in the last 15 years. The project is likely to have grave effects in the Bay of Bengal where the Ganges meets the ocean, as already there are dead zone pockets in the Bay.

Dead-sea zones do not always produce similar effects, but one near the deltaic region could be catastrophic for the Sundarban region. The region is already threatened by large discharges of untreated sewage, estimated to be 680 million tons per annum, including hazardous industrial waste and fertilizers. Reduced flows in the Ganges would only exacerbate the situation. In spite of the $33.3 million spent in the 1980s on cleanup, the pollution levels continue to be alarmingly high with serious consequences for all, including humans.[17]

In the Gangetic systems the decreasing flow of upstream water is increasingly overwhelmed by the saline tidal flows in the lower stretches of the Ganges, with weak freshwater flow unable to resist the force of seawater during tides. It is a major concern for the greater Kolkata region, which accommodates some 18 million people. The Ganges supplies the municipal and industrial waters for the region. The UN Education, Scientific and Cultural Organization (UNESCO) identified Kolkata as one of the five mega-cities in the sub-continent under severe water-stressed conditions. It expressed concern about unsustainable abstraction rates; reduction in the dry weather base-flow in some downstream watercourses; irreversible aquifer damage due to increasing saline intrusion; and damage to groundwater-dependent ecosystems (UNWWDR 2003: 160, 81). Further reductions in flow from the upstream will be catastrophic for the region.

The delta and the Sundarban forest did not grow in one day—or even in one month, year, or century. As the chemistry and physiology of the delta changes, so do their effects on the formation of mangroves, on animal life, and on the estuary fisheries. The effects on the world heritage–listed Sundarban forests, the largest of their type in the world and the home of the nearly extinct Royal Bengal tiger, also need to be examined.

Is the Timetable Realistic?

The government planned to complete the river-linking project by the end of 2016. It had listed five distinct sets of activities to be completed by 2006. The premises on which its timetable is based have not been made clear. The schedule of activities, including specific tasks, are completion of feasibility studies; estimation of costs; identification of concrete benefits; consideration of options for funding; and devising methods of cost recovery. Nowhere in the schedule can the word *environment* be found.[18] And, unfortunately, beyond a few lines drawn on the map, no scientific and technical information has been made available to the public (Bandyopadhyay and Perveen 2004: 5307–16).

Even if it is assumed the project is viable, the time allocated for finalizing the project is scandalously deficient. Yet the timetable is critical, as enough time must be available for the government, academics, scientists, and appropriate agencies to examine the project's immense consequences. The GoI's timetable to complete it defies logic. The question needs to be asked why there is such a hurry, particularly when so many issues still need to be scrutinized and evaluated before a final decision can be made. A few experts have already questioned the attitude of secrecy on the part of GoI, have judged the practice dangerous, and have argued that there is no need to hurry and that the GoI must allow sufficient time to evaluate the project in the fullness of time (Mukherjee 2006).

International experience shows that projects of lesser magnitude take significantly longer to design, evaluate, execute, and complete, if everything is done properly. Broadly speaking, large-scale projects related to major rivers or river valleys need a period of about fifty years for design, construction, and use (Crow 1995: 188). For example, the Snowy Mountain project in Australia, an engineering marvel in its own right, which involved only two states within the federation plus the federal government and no riparian country, took some fifty-five years to complete (Raymond 1999: 8–19). By any standard of imagination, the river-linking project is a mega-project, the equivalent of which is almost certainly unheard of anywhere in the world—and yet the government is rushing with extraordinary speed.

Notes

1. When rainfall in a region is more than 25 percent below normal.
2. See http://nwda.gov.in/index2.asp?sublinkid=62 (January 19, 2006).
3. *Statesman* (2007), "River–link agency seeks Rs 241 cr.," February 9.

4. *Times of India* (2003), March 17; *Statesman* (2003), August 20.

5. The eleven challenges to satisfy human life and well-being are: Basic Needs and the Right to Health; Protecting Ecosystems for People and Planet; Cities: Competing Needs in an Urban Environment; Securing Food for Growing World Population; Promoting Cleaner Industry for Everyone's Benefit; Developing Energy to Meet Development Needs; Mitigating Risk and Coping with Uncertainty; Sharing Water: Defining a Common Interest; Recognizing and Valuing the Many Facets of Water; Ensuring the Knowledge Base: a Collective Responsibility; and Governing Water for Sustainable Development.

6. Phil Williams, President, International River Network, an internationally well-known pro-dam engineer turned anti-dam crusader, and Ernest Razvan of the International Hydraulics Institute at the ICOLD Vienna conference in 1991 (Pearce 1992:141, 348).

7. *Statesman* (2004), November 13; BBC News (2003), August 28; *Guardian* (2003), July 24.

8. *Economic and Political Weekly* 38/40, pp. 4277–4280.

9. "Spain ditches Ebro river project"; http://news.bbc.co.uk/2/hi/europe/3817673.stm (June 19, 2004).

10. *Anadabazar* (2003), "Nadi Jude-deo-ar Prakalpe Samuha Bipad Paschimbanger" [Bengali vernacular], February 25; and "Indian river project would be 'catastrophic' for Bangladesh"—Bangladesh Minister; http://story.news.yahoo.com/news?tmpl=story&cid=1534&ncid=1534&e=6&u=/afp/ (December 18, 2004).

11. At the time of public hearing for the EIS of the Sethusamudram Ship Canal project the people directly affected by it were denied access before the public hearing at the behest of senior bureaucrats of the project; *Statesman Weekly*, October 30, 2004.

12. Before the Aswan Dam was constructed, the Nile carried an average of some 124 million tons of sediment to sea each year and deposited another 9.5 million tons or so on the narrow flood plain and delta, home to almost all of Egypt's population. After a quarter of a century, it was possible to see the long-term consequences: besides an enormous increase in the amount of sedimentation in the lake, there are also other severe damaging consequences (Pearce 1992: 285; Lavergne 1986; Ward 1997: 51). The Mississippi delta is also creating a different type of problem that is no less critical than that of the Nile delta (McCully 1996: 34–35). An independent Canadian study of the Three Gorges Dam in China claims that there was no realistic way the reservoir could be managed, as it fills with sediment and loses its flood-control storage, intended to protect millions of people who have been induced to move into flood-prone areas downstream (Williams 1993).

13. The Gangetic Delta covers two countries: India and Bangladesh. It is the most active delta on earth. It deposits silt 2,000 nautical miles away in the Bay of Bengal (King 1983: 119).

14. British Broadcasting Corporation (2007), "Feeding the World"; the BBC World Service series investigates the growing but often underreported challenges facing the world's food supply. The reporter went to the Ghoramara Island in the Gangetic Delta

in Sundarbans (India) from where he filed this report following his discussions with the long-time residents (Growing Pains in Part One); http://news.bbc.co.uk/2/hi/programmes/documentary_archive/6500041.stm (March 29, 2007).

15. *Statesman* (2005), "Walls cannot stop waves, mangroves can," January 28: www.thestatesman.net/page.news.php?clid=2&theme=&usrsess=1&id=66880; and Abraham (2005).

16. About 30 million m3 of sand would have washed into the Cambridge gulf and dispersed along the coast in large floods, if the Ord river dam had not been built. Since then there have been few large floods capable of flushing the silt. The dam has robbed the pristine coast of natural sand drift and contributed to the beginnings of a dangerous dieback of the mangrove nature reserve at the mouth of the Ord. This mangrove forest is Australia's largest and a crucial nursery for the coastal prawn fishery; *Australian* (March 19, 2004).

17. *Statesman* (2004), School of Oceanographic Studies, Jadavpur University, August 7; *Statesman* (2005), "Ground Reality Gets Out of Control," April 6.

18. Timetable for interlinking rivers: [i] Notification of the Task Force: 16 December 2002; [ii] Preparation of Action Plan—I, giving an outline of the time schedules for the completion of the feasibility studies, detailed project reports, estimated costs, implementation schedule, concrete benefits and advantages of the project, etc.: 30 April 2003; [iii] Preparation of Action Plan II, giving alternative options for funding and execution of the project as also the suggested methods for cost recovery: 31 July 2003; [iv] Meeting with the Chief Ministers to deliberate over the project and to elicit their cooperation: May/June 2003; [v] Completion of Feasibility Studies—already in progress: 31 December 2005; [vi] Completion of Detailed Project Reports [DPR]—Preparation of the DPRs will start simultaneously since Financial Statements [FS] in respect of six river links have already been completed : 31 December 2006; [vi] Implementation of the Project: 31 December 2016.

PART TWO

ISSUES FOR INDIA AND THE REGION

CHAPTER SIX

Environmental Perspectives

In the new millennium, development of *water policy* cannot be undertaken without in-depth assessment of the possible impacts of climate change on physical environment, rainfall, economic, governance and security issues, and other related aspects. Scientists, environmentalists, and others have been focusing on these issues for some time now (Gelbspan 1997: 8–10 and 177), but some argued to the effect that "the jury was still out" and refused to seriously bring these issues within the purview of their respective environmental, economic, and social policies, and pointed out the uncertainties of estimates of climate change. No doubt there had existed uncertainties (Dickinson 1989: 5–13), but the situation has significantly changed in recent years.

The Intergovernmental Panel on Climate Change (IPCC), jointly established by the World Meteorological Organization (WMO) and the UN Environment Program (UNEP) to assess information related to climate change, its potential impacts, and options for mitigation and adaptation, at its meeting in January 2007 attended by 2,000 of the world's leading climate scientists, unanimously concluded that "there was now little doubt that human activity was changing the face of the planet." The Panel agreed that the rising surface temperatures around the world were linked to man-made emissions of greenhouse gases, to significant increases in ocean temperatures, to rises in sea levels, and to the dramatic melting of Arctic sea ice over the past thirty-five years (Connor 2007; Borenstein 2007). This report by the Panel went further than any of its previous reports.

The UN's Framework Convention on Climate Change (UNFCCC), which relies on the IPCC scientists, came into force in 1994 after being formally approved by over 100 countries. Following this a process called Council of Parties (CoP) was established to negotiate responsibilities for action on climate change as part of the governance of the Framework. The third CoP, held in Kyoto in Japan, established what is now known as the Kyoto Protocol, an international plan designed to reduce climate change pollution. India is a signatory and has ratified the Protocol, but is not required to meet any specific gas-reduction targets yet, although during 1990–2000 carbon dioxide emissions increased by 69 percent in India.

The IPCC launched its latest assessment of present and future impacts of climate change in Brussels in April of 2007 that was endorsed by the representatives of the member-country governments. The assessment emphasized that time was running out for the member countries, if catastrophic consequences were to be avoided in the very near future. It further argued that in such circumstances the world's poorest of the poor will be hit hardest. Following this assessment the UNFCCC has issued a statement that the world urgently needs to launch an agreement on future international action to combat climate change as well as to look effective ways to generate funds needed for future adaptation.

The existing targets for countries set by the Kyoto Protocol will be renegotiated effective 2012, and many countries, particularly the EU, Canada, and Japan have already indicated that the two exempted countries (India and China) will be required to meet specific targets then. The compliance costs and its economic consequences are likely to be significant for India, as India is considered a large polluter along with China. Globally, debate is already heating up as to whether developing or developed nations should bear the cost. Although fairness demands that rich countries who have contributed most in creating the problem in the first instance should pay to rectify the situation, realities remain that international pressures from developed countries are mounting that two of the fastest growing economies must also contribute their reasonable share of the costs to minimize atmospheric greenhouse gases. One suggestion gaining some currency is that the rich countries contribute to transition costs, but no nation can escape responsibility, and all should take action to combat global warming (Rosenthal and Revkin 2007). For India, a resource strained country, demand on her limited financial resources could be significant.

The climate change is caused by the burning of fossil fuels like coal, oil, and natural gas, which releases carbon dioxide and other pollutants into the atmosphere. The affect of climate change is felt in a number of areas. Some of the critical areas are:

- Affect on annual runoff, runoff variability, and seasonal runoff;
- Runoffs, in turn, affect water supply, agricultural production, flood-management, and hydropower generation;
- Impact on watershed, increasing soil erosion and altering the hydrological response of the watershed.

These affect water supply and also impact on the entire riparian ecosystems (Williams 1989: 83–93). Besides these direct consequences, enormous indirect impact is also felt on three other critical areas, namely, economy, health, and national security. These impacts are not felt uniformly (Bardach 1989: 117–50; Vellinga and Leatherman 1989: 175–89). For example, the *Gangotri Glacier* is receding rapidly and many Himalayan glaciers are not gathering much snow, as has been the case in Canada and the European Alps, but its impact hardly will be felt in the western or southern India. A few of these issues are discussed earlier (chapter 4).

In practical policy terms, countries are grappling with how to address these issues. Of many governments, the California state government in the United States has already earned global appreciation for its initiative to combat environmental consequences of the "greenhouse effect." At the informational hearing on 'Climate Change and Water Resources' of the Californian Assembly Committee on Water, Parks and Wildlife submission made by the Pacific Institute (Gleick 2007) succinctly articulated how to approach this issue in clear terms. The concerns are both real and urgent, and all stakeholders need to be roped in to ensure that people understand the issues clearly to ensure policy success. The Institute's submission reinforced the importance of articulating the issues in simple terms. This is the surest way to get stakeholders involved in addressing an issue that is critical for everybody, namely, access to freshwater. Four aspects highlighted by the institute are identifying:

- What has been done to deal with water related risks to climate change;
- What has not been done;
- What should not be done; and
- What should be done.

The river-linking project plans to transfer the water held in dams in the Himalayan Rivers during the rainy season to the Peninsular Rivers through gravitational processes or pumping. Although the latter will require significant energy input, no reference has been made to its environmental and economic costs. The American experience suggests that the cost of delivering water by pumping could be enormous, both environmentally and economically

(Natural Resources Defense Council 2004). For example, the energy required for taking water to Karnataka, the driest southern state, would be great indeed, because many parts of the state are 600 meters above sea level.

Dams interfere with nature. The degree of interference is determined by their size, type, and other factors. There are seven ways that a dam can impact on a river basin:

- It can cause upstream change from river valley to reservoirs;
- It can produce changes in the downstream morphology of river beds and river banks, deltas, estuaries, and coastlines due to altered sediment load;
- It can change downstream water quality;
- It can reduce biodiversity due to the blocking of the movement of organisms and the changes created by it;
- It can cause changes in total flows, such as in the seasonal timing of flows, and short-term fluctuation in flows, including extremely high and low flows;
- It can cause changes in downstream water morphology and water quality caused by the altered flow patterns; and
- It can reduce riverine, riparian, and flood plain habitat diversity (McCully 1996: 30–33).

Few studies have been done on the impact of large reservoirs on the environment. But the results of studies carried out by the World Bank indicate that the adverse impact of large dams on climate and vegetation could be significant. Furthermore the IPCC noted that the increased runoff induced by climate change could pose a severe threat to the safety of existing dams with design deficiencies, if nothing else, and that these dams may require reevaluation to incorporate the effects of climate change (World Meterological Organization 1991: 181; Legett 1995, cited in McCully 1996: 145).

Water storage and changes in water flows affect climate, weather, flora and fauna, community health, and the economy. To achieve a balance between the environment and the economy, the effects of water storage and changes in water flows on the environment need to be assessed (Cumberland and Herzog Jr. 1977: 21). These need to be examined in a transparent way so as to make clear that the project is not meant to harm the interests of riparian countries in the river basin or the affected states. A clear assessment of the proposal would help prevent people from coming out of the woodwork to extract political mileage from it, as even a small problem often gets complicated in the sub-continent.

Climate Change and the
Environmental Impact Statement (EIS)

Climate change is a critical issue in water policy, as the effects are not geography-specific and it is not easy to describe the impact of environmental laws in spatial terms. Some environmental issues may need to be considered in a wider biospheric context (Kiss 2002: 131).[1] The importance of climate change is clear from the fact that the warmest year at the world's surface since record-keeping began in the 1860s was 1998; followed by 2002, 2003, and 2004. The Indian experience is not fundamentally different from the global trend.

The IPCC estimates that, globally, carbon dioxide concentration will rise by 450 to 550 parts per million (PPM) by 2050, causing a temperature rise of 0.5 to 2.5° C, and a rise of 1.4 to possibly 5.8° C by the end of the century (Browne 2004: 20–32).[2] The World Wide Fund for Nature (WWF) found that in the last 100 years, northern Europe has become 10 to 40 percent wetter and southern Europe up to 20 percent drier. It also found significant changes in plant and animal behaviour in the Netherlands and southern Europe. The European Environment Agency (EEA) argued that global warming has been evident for years; what is most troubling is the speed of recent changes (Kelly 2003). Three recent publications (Flannery 2005: 4; Hansen, Nazarenko, Ruedy, Sato, Wills, Genio, et al. 2005: 1341–45; and Kolbert 2006[3]) have left no doubt that the climate situation is deteriorating rapidly and urgent steps are required to be taken by the global community before a point of no return is reached.

India signed the Kyoto Protocol in 2002. The primary concern over climate change must be the level of investment required to overcome its consequences (see Stern 2006). Even if corrective measures are adopted at the global level, India's diverse physical features and vast size will make it imperative to undertake local initiatives to counteract the diseconomies caused by climate change in order to remain globally competitive, as environmental problems will become manifest in different ways in different regions (Kütting 2004: 132).

Past experience shows that an alliance of bureaucrats and the construction industry has maintained a tight grip on the irrigation industry and through it on India's water policy. They have failed to take even the minimum steps required to ensure that environmental concerns are considered in water management policies and projects under their control (Mollinga and Bolding 2004: 5 and 291). Their strength should not be underestimated. They maintain their power base through funding water projects in a way that some researchers have aptly called *project culture*, which is structured and defined by past events and agendas and carries with it a legacy of outdated goals and interests (Nickum and Greenstadt 1998: 141–61; Donahue and Johnston 1998: 343). India needs to tackle this problem head on, as it will not be possible for the nation to buy itself out of these difficulties as richer countries do.

Greenhouse gas emissions harm people in developing countries most, especially the marginalized and the disadvantaged (Kelly 2004; Altman 2006). In 2003, there were 700 floods, storms, and other weather calamities that claimed 75,000 lives worldwide and caused about $65 billion in damage.[4] About 98 percent of the victims were in the poorest nations. Loss of life and damage to property in industrialized nations was much lower, as they use weather predictions and prepare for the calamities.[5] For example, the 2004 cyclone devastation in Haiti also affected the neighbouring Dominican Republic, which is better off, but caused much less economic devastation there.

Various independent studies confirm the disastrous consequences of climate change on the global physical environment and flora and fauna. The most serious of all environmental threats is the melting of the ice caps. The Arctic Climate Impact Assessment Group (ACIAG)[6] found that the Arctic ice cap has shrunk by 15–20 percent in the past 30 years and that the Arctic temperatures are increasing by almost twice the UN-projected global average. The Arctic Ocean could become almost ice-free in summer by the end of the century, and the melting on Greenland alone could add 10 centimeters to global sea levels by 2100. The effects would be felt all over the world: from Bangladesh to Florida to Tuvalu in the Pacific.

Other glaciers such as the Athabasca in Alberta retreated by 1.5 KM in seventy years and lost 16 million CM of water per year; that is more water than the Athabasca glacier regained from winter snows. The Peace River, the largest of the western Rockies' rivers, now has flows that are 60 percent of those of the mid-twentieth century. Lower snowfalls and the construction of the Bennett Dam upstream are believed to be the causes. The consequences for the vast Peace-Athabasca Delta, the largest freshwater delta in western North America, are disastrous (Schindler 2003). The Swiss glaciers shrank a meager one percent in a twelve-year period to 1985, but lost some 18 percent of their area in a fifteen-year period to 2000. The Gangotri Glacier that feeds the Ganges has already receded by 14 KM in the last 100 years. Japanese researchers also found that many glaciers have retreated by between 30 and 60 meters during the last three decades.

The World Conservation Union (WCU) found considerable anecdotal evidence that globally the number of amphibians is declining at an unprecedented rate and that in some cases they are moving towards extinction. It also found that even under the most favourable assumptions, the coral cover in the Great Barrier Reef in Australia will cover less than 5 percent of most reefs by the middle of the century; there is little to no evidence that coral could adapt fast enough to match even the lower projected temperature rise. Other earlier studies also predicted that high sea-water temperature could lead to the death of coral reefs (Brown and Ogden 1993: 64–71). Norwegian scientists found that the release of huge amounts of hydrocarbon in the seabed could lead to a massive soaring of the earth's temperature by 5 to 10 degrees Celsius, a situation that previously arose about 55 million years ago (Dickens 2004: 513; Svensen, Planke, Malthe-Sorennsen, Jamtveit, Eidem, et al. 2004: 542–45).

Even scientists from countries such as China, previously ambivalent about climate change, are shifting from their earlier positions. A team of Chinese and the US scientists found that during the last forty years glaciers in western China's Qinghai-Tibet plateau have been melting by 7 percent annually and that the shrinkage has worsened since the early 1990s. China's chief glaciologist warned that most glaciers in the region could melt by 2010 if global warming continues.[7] Tibet is the source of most of Asia's mighty rivers, including the mighty Brahmaputra and the Indus.

The melting of the Himalayan glaciers is of great concern to India. The reasons are simple: Nepal has about 3,300 glaciers of which about 2,300 contain glacier lakes. Between 70 and 90 percent of the water discharges to the Ganges are from the Nepalese rivers and lakes, and if glaciers dry up so will the Ganges downstream. In the initial stages these lakes can also breach their shores causing enormous damage downstream (Dhakal 2003). Scientists agree with such predictions, though a lack of reliable data prevents them from determining precisely how many of them are close to breaching, which could cause enormous disaster to lower-riparian countries.

EISs May Not Provide All the Answers

The existing methodologies used in preparing EISs are far from satisfactory, and vested interests have often taken advantage of this to achieve their objectives by making unrealistic assumptions. While such manipulations are prevalent in developing countries, industrialized nations are not immune from them either. For example, the Teton Dam in the United States breached within one year of its opening in the nineteen seventies because designers and engineers assumed a much lower water flow than actually occurred (Reisner 1986: 398–424). The reasons for such practices are not hard to imagine:

- The majority of the post-war dams in developing countries were funded by either multilateral agencies or bilateral donors, and more likely than not were also designed, implemented, and evaluated by foreign agencies using foreign equipment procured in foreign markets (Hancock 1989: 155). Some 85 percent of the untied Japanese loans to the poorest countries were spent in Japan (Forrest 1991: 24–32).[8] And at least one-quarter of the $60 billion in loans and grants by major donor countries were tied grants or loans (Randel and German 1994); and
- International environmental consulting is now a very profitable business for a few industrialized countries. More worrying is the absence of any review of the quality of the consultants' work and the total absence of any control mechanisms. Sixty percent of the national dam agencies acknowledged in 1991 that there was no formal system for monitoring the impacts of dams in operation, despite the claim in almost every EIS that environmental monitoring would be a key part of the follow-up (McCully has clarified this issue further [1996: 54–58]).

The task of preparing an EIS is complex, and it becomes even more complex when not much information is available. In the sub-continent the task becomes even more complicated, as data developed by one country on hydrological conditions, water availability, and the effects on dry-weather flow are often questioned by other riparian countries when it suits them to do so (Kayastha 2001). While it is difficult to get all the answers from an EIS, it at least identifies issues, which need to be considered before taking a final decision on a project. If nothing else it at least reduces the scope for major mistakes in the downstream of projects. India ignored this aspect of project formulation at its peril in the past which is reflected in her inability to reap the benefits from her huge investments in the water sector.

A generalized template for an EIS for projects is provided in Appendix Two.

Notes

1. Also, see *NY Times* (2004), August 17.
2. Griggs (Hadley Centre for Climate Prediction and Research), Meehl (National Centre for Atmospheric Research), Howden (CSIRO, Australia), and Carter (James Cook and Adelaide Universities) also expressed similar views. See www.abc.net.au/rn/talks/brkfast/stories/S1138130.htm (June 24, 2004).

3. The *Economist* (London) has reviewed Kolbert's book along with Flannery's book. The reviewer concludes, "Both books provide the same central message: act now. Every year's delay in doing something about climate change will take far more than a year to put right. Once the ice is gone, it will not come back. Once the permafrost melts and the methane it contains is released, it cannot be recalled—and methane is far more potent greenhouse gas than carbon dioxide" (*Economist*: March 4, 2006; p. 93).

4. *Guardian Weekly* (2001): the 2001 IFRCRCS Report found that before 1996 disasters caused by landslides, floods, and storms numbered about 200 p/a and had risen to 392 in 2000; July 5–11. Also see *NY Times* (May 18, 2004).

5. Developing countries often lacked preparedness plans, partly from lack of funds and experience, but also from lack of political will, said the Executive Director of the American Meteorological Society (Olson 2004). However, experience shows that even industrialized countries on occasion are exposed to disastrous consequences, e.g., the disaster caused by the hurricane *Katrina* in New Orleans, USA, in 2005; Wikipedia, *Hurricane Katrina*; http://en.wikipedia.org/wiki/Hurricane_katrina (February 4, 2007).

6. The United States, Russia, Canada, Denmark, Norway, Sweden, Finland, and Iceland are all members of the ACIAG.

7. *China Daily* (2004), www.abc.net.au/news/newsitems/200410/s1213705.htm; October 6.

8. Also see *Economist* (1993), "H-Street Blues"; May 1; pp. 83–84.

CHAPTER SEVEN

Economic and Financial Perspectives

The river-linking project is estimated to cost about $125 billion, consisting of $23.6 billion for the Peninsular part, $41.1 billion for the Himalayan part, and $60 billion for the hydro-electricity component (Shrivastava 2006: 44). The government considers this is a very rough estimate, as the executive chairman of the government-appointed Task Force has indicated the cost could reach $200 billion (Shiva 2003: 73).[1] Given the size of the cost in relation to India's gross domestic product (GDP: $510.2 billion in 2002—UNDP 2004: 186) and that the project timetable requires expending these funds within a relatively short time, financial and budgetary related issues need careful examination to assess the project's fiscal and economic implications. The total cost, therefore, could be anything between 25 to 39 percent of the 2002 GDP of the country. Details of the cost breakdown and the bases of the estimates are not yet available for public scrutiny.

With such a mammoth and complex project, cost overruns are inevitable. Past cost overruns in major projects in India varied between 49 and 147 percent (World Bank 1991: 9–10; Rich 1994: 98). Thus the eventual cost could reach between $185 billion and $306 billion, assuming that the inflation rate remains moderate. Inflation, based on the wholesale price index, has been hovering around 7 percent in recent years (Panda 2002). In August 2004 it increased to 8.74 percent. Many economic commentators also argue that the GoI-published inflation rate cannot be believed.[2] More worrisome is the experience of the recent-plan period that witnessed project delays and experienced cost overruns of between 50 and 893 percent in a few of the large water development projects (Rath 2003: 3032–33; Singh 2004).

116   Chapter Seven

If similar delay is repeated, which is possible because of the project's complexity, the total cost could even reach beyond one thousand billion dollars. Such cost explosion is unlikely, but this scenario needs to be considered within the country's overall fiscal condition. In this context it is also worth remembering that the costs of dam construction have been steadily rising over the years.[3] This is reflected in the rising costs per hectare of new irrigation in India by almost 60 percent in real terms between 1979 and 1985. Since then the situation has not improved; if anything, it has deteriorated significantly.

These estimates were based on the assumption that the inflation rate remain moderate around 2002 levels. Given India's dependency on imported energy resources, energy's contribution alone in the future inflation rate could be substantial. The global oil price has been hovering between $70 and $80 per barrel in 2006 and was hovering around $100 in 2007. Currently, global oil production is about 82 million barrels a day, but rising demands from China along with other uncertainties are causing oil prices to soar. Supply is falling behind demand, and some net exporters of energy are now becoming net importers (e.g., Indonesia). India imports about 75 percent of its energy needs and the demand is increasing exponentially. The government's 2006–2007 Economic Survey has also expressed concern on volatile oil prices that may put additional pressure on the inflation rate.

The XI Finance Commission found that the fiscal profile of the country was perhaps worse than ever before, with almost every key fiscal variable moving in a negative direction. All states had suffered large revenue and fiscal deficits that were showing no sign of improvement.[4] The 2001–2002 economic survey also found that the domestic gross savings of the country was even lower than its estimated cost of the project (Singh 2006: 16).

The River-Linking Project in Sectoral Developmental Context

Countries that have achieved significant economic growth during the last few decades invested heavily in two sectors, namely, basic education and health. India did not, although other developing countries in similar situations, such as China, are doing just the same. For example, in 2001 China's public expenditure on health was 2 percent of GDP against India's 0.9 percent. On the basis of PPP, China's per capita health expenditure was $224, compared to India's $80. There is also a growing concern about HIV infection in India. UNICEF has expressed concern about India's failure to fulfil its commitments to combat HIV-AIDS and has stressed the need for effective measures to safeguard children from pandemic infection.

In 2003, India's HIV prevalence rate was 0.4 to 1.3 percent. The rate in developing countries as a whole was 1 to 1.6 percent. It was 0.3 to 1 percent in South Asia, 0.2 to 0.5 percent in OECD, and 0.1 to 0.2 percent in China (UNDP 2004: 157–58; 165–67). The 2006 UNAIDS Report finds that India has overtaken South Africa as the country with the largest number of people living with HIV virus. The report estimated 5.7 million Indians were infected by the end of 2005, causing huge loss of life and loss to the national economy (assuming the 2005 Asian death rate of HIV patients, some 357,000 of these people would have died in 2006—UNAIDS 2006; Mitra 2006). It is virtually impossibly to predict the level of expenditure that will be needed to successfully combat HIV, but certainly it will be very large.

India's failure to achieve a major reduction in child mortality rates is another health concern. Its reduction was even lower than those of Bangladesh and Nepal. During 1990–2001 Bangladesh and Nepal brought their child mortality rates down by 46 and 37 percent, respectively, compared to India's 24 percent. China also achieved significant success: its rate was 39 per thousand children, compared to India's 93. To reduce the rate further the World Bank suggests making greater investment in very young children and targeting public financing for poor children.[5]

Another emerging health-related issue is aging. The GoI has yet to consider aging within broader educational, economic, social, and health-policy contexts. And the practice of managing aging within the family is rapidly breaking down (Lloyd-Sherlock 2004). In India the state of Kerala had the largest percentage of people aged sixty-five and over in 1991, and the state of Assam had the lowest. The preliminary 2001 census data show that Kerala had achieved a 91 percent literacy rate compared to Assam's 53 percent (which is much lower than the Indian average of 65 percent)[6]—confirming a direct link between education and longevity. Public expenditure on education and health are positively correlated. If the issue of aging is not handled effectively it could trigger a much more serious crisis that could eventually threaten democratic forms of government (Petersen 1999: 42–55).

In India public expenditure in the education sector as a whole has increased marginally over the last decade (1990–2000) as expenditure on the primary education sector has declined proportionately. In 1990, the share of the primary education sector in all public education expenditure was 38.9 percent, and in 1999–2001 it declined to 38.4 percent (UNDP 2004: 175). This indicates a need for additional investment in education (Gulati and

Rajan 1999: WS46–WS51), as economists agree that high-quality primary education is a critical element in achieving rapid economic growth. The spectacular economic growth in Taiwan, South Korea, and China has been due in large part to high investment in the primary education sector that prepares the base for a high-quality work force that can cope with rapidly changing technology (Bhagwati 1996: 9; Sen 1997). An economist (Basu 1997) aptly described India's unbalanced education structure in the following terms: "for most countries, the [education] house will look like a pyramid, tapering off as one goes higher up. In the case of India, the building is more like a tower."

If one looks at public health and education expenditures in the budgetary context, India's position stands out as far worse than many of the "low human development nations." The 2006–2007 Economic Survey acknowledged that India has a very high rate of illiteracy and infant mortality rates. This is reflected in India's Human Development Index ranking: in 2004 India's global ranking was 124, and that slipped to 126 in 2006. Most South Asian Association of Regional Cooperation (SAARC) countries and China spend more public funds on health than India. In 2006 the Union health ministry conducted a national family health survey. The survey finding speaks for itself: "Despite the excellent growth record for which India is getting appreciation, its social indicators are dragging mainly due to poor access to health care."[7] More troubling, however, about India is the lack of any real political will to ensure the effective use of whatever resources are committed to the health and education sectors.

The government has also experienced additional policy-pressure from the National Human Rights Commission (NHRC), which argued that "health was one of the basic Human Rights" and hence, to overcome the non-availability of doctors in the rural areas the government should consider making rural posting of doctors compulsory before doctors getting registered with the Medical Council of India.[8] Most certainly inter alia this will require huge financial commitment from the public exchequer. It is also worth remembering that in absolute terms, many of the African and Central and South American countries commit more funds for primary education and primary health care, even though these countries are forced to spend more on debt servicing.

A Weakening Fiscal Environment

Collectively the GoI's and the states' fiscal position is weak. In fact, the fiscal positions of the Indian states are getting worse. Unsustainable budget deficits, poor performance of public undertakings, and unsustainable increases in salary and pension payments are all part of the picture. The combined gross fiscal deficit of the center and the states in 2004–2005 exceeded budgeted levels, but the increase in combined gross fiscal deficit of the states outpaced the rise in receipts, resulting in a higher fiscal deficit.[9] While the federal fiscal deficit was 4.8 percent in 2003–2004, state governments collectively accounted for 40 percent of the combined deficit. Considerable belt-tightening would be required to achieve the target of reducing the deficit by 0.3 percentage points each year under the Fiscal Responsibility and Budget Management Law (FRBML). The 2005–2006 Central Budget planned for only a 0.2 percentage point reduction.

Even if it is assumed that the GoI is able to achieve its fiscal targets, which is most unlikely, what about the states? Of the twenty-five states, the debts of only five were between 10 and 20 percent of their respective GDPs; those of eight states were between 20 and 30 percent; another eight states' debts were in the 30 to 50 percent range; and the remaining four states' debts were above 50 percent of their respective GDPs. States' outstanding combined long-term debts increased by about 120 percent during the 1995–2000 period. In one year alone (1999–2000), their combined debts increased by a little more than 20 percent. Also, most states pay a very large percentage of their revenue towards interest payments: six major states pay more than 20 percent towards interest payments on their debt; only four small states pay less than 10 percent; and the remaining states pay between 10 and 20 percent.[10] Annual interest payments of the aggregate central and state loans now amount to about Indian Rs.2 trillions against a public debt of about 30 trillions.[11]

More importantly, economic commentators have now started questioning the government's deficit figures and have argued that the actual difference between government expenditure and income has been made to appear much smaller than it really is. One commentator (Roy 2007) argues that "the significance of this discrepancy will not be lost on anyone seriously concerned to address India's fiscal and monetary problems."

Furthermore, many state governments are already financially bankrupt. Some even fail to make salary payments on time and most states and local governments spend more than 50 percent of their revenue to pay their staff salaries. Quasigovernment employees in some states have not received their salaries for many months, and in 2005 the Supreme Court had to issue orders directing one state government to make pro-rata payments to staff in one of the eastern states.

After the fifth Pay Commission Report for central government employees in 1997, the GoI predicted that in some states the salary payments could go up to as high as 95 percent of revenue (Mukarji 1997: 750–53). The Finance Commission then suggested that future pay commissions may be few and far between and should, among other things, take a comprehensive and normative view of the finances of the center and the states.[12] The Commission also concluded unequivocally that government's fiscal situation has worsened and warned about its serious consequences for the future.

Unfunded pension liabilities will also have enormous effects on government finances. During the 1988–1997 decade the combined revenue expenditures of the center and the states increased by 201 percent, compared to 235 percent in administrative services expenditures and 284 percent in pension payments for public servants and allied workers (World Bank 1997a: 92-3). In this area the GoI has at least made an attempt to reduce the future burden by introducing a Contributory Pension Scheme. Nine states are on the verge of signing up for the scheme, but some are refusing to do so.[13]

It is of further concern that coalition governments affect the governments' fiscal policy choices (Khemani 2004: 125–54). Many Indian states are now administered by fragmented coalition governments, including the central government. A study of the composition of 16 Indian state governments covering a period of twenty-one years found that only three of those states did not have a coalition government during this period, although the duration of coalition governments in states varied (Chaudhuri and Dasgupta 2006: 645).

How to Fund the Project?

It is not clear how the government intends to fund the project. From the available anecdotal evidence it appears that the government is thinking in terms of, firstly, imposing additional irrigation cess, or taxes; and secondly, of encouraging investment, particularly direct foreign investments that would help boost economic growth and generate increased revenue (growth dividend). The question is, how realistic are these options?

It defies logic how such a large funding requirement can be met through additional irrigation cess. One needs to keep in mind that the economic and financial returns on investments in dams in India during the last five decades have been meager. The World Bank found that of the nine large irrigation projects completed in 1989, only two had economic returns above the estimated opportunity cost of capital (Rich 1994: 98). Many irrigation projects have been found to not even cover the running costs, let alone the capital costs.

A major problem is that in spite of high growth rates in recent years, the agricultural sector's performance has been poor. The 2006–2007 Economic

Survey acknowledged buoyancy in all sectors of the economy except agriculture, which was languishing with a growth rate of 2.7 percent only. The 2002 National Sample Survey (NSS) found that the average value of household liabilities in 2002 was $9.82 per rural household and $7.35 in urban areas. More importantly, the value of liabilities had increased in rural areas since the 1991 survey. About 73 percent of households in India were in rural areas; 60 percent of them were headed by cultivators. Given that most farmers cannot even pay for irrigation water and rural household debts are increasing, how can the policy makers expect to collect additional irrigation cess?

On the social front, unemployment and poverty remain hidden time bombs and are feeding the discontent simmering in many parts of the country (Lewis 1995: 188). Recently published official data reveals that the total employment in the organized sector fell from 28 million to 27 million in 2003 and to 26.4 million in 2004. While employment exchange registered "unemployed data" are notoriously inaccurate, it provides some indication of the unemployment situation in the country. Employment exchange data show a dramatic rise in educated and uneducated youth in the fifteen to nineteen age group. In 2000 the number of registered unemployed in this age group was about 20.9 million that increased to about 30 million in 2001 and to about 40 million in December 2006.[14]

Santarelli and Figini (2004) found that foreign direct investment has consistently had no significant effect on poverty alleviation in developing countries. This was also confirmed by Vivarelli (2004), who found that foreign direct investment inflows do not show any significant distributional effect when other determinants of in-country income inequality are controlled for. The UNDP and ILO (2007) in their joint study of *Asian Experience on Growth, Employment and Poverty* concluded that employment growth has not matched the level of economic growth, and "while this has been the case, we also see the region (that includes India) facing a continuing decline in employment growth."[15]

More importantly, future employment growth prospects also remain grim. The combined organized sector annual employment growth rate declined during 1994–2004 period from 1.20 percent to 0.38 percent, although it recorded some growth in the private sector during the same period from 0.44 percent to 0.61 percent. However, the biggest concern is the widening gap between the annual labour force growth rate and employment growth. The 2004–2005 NSS found that during the 1999–2000 and 2004–2005 periods the labour force grew annually by 2.54 percent compared to an employment growth rate of 2.48 percent. This was reflected in an unemployment rate of 3.06 percent in 2004–2005, compared to 2.78 percent in 1999–2000.[16] Furthermore, the government's

2006–2007 Economic Survey acknowledged that India has one of the highest rates of growth of working age population among the more populous countries of the world, no scope for policy complacence.

These clearly indicate that the government will have to address the labour-market issue with utmost urgency if social-unrest and major social-dislocation is to be avoided. A large number of pockets with simmering unrest has been a common feature all over the country during last few decades, and governments of all persuasions can ignore this situation only at their own peril. Labour market reform to create job opportunities for millions of unemployed will require both innovative policies as well as capital investments in education, vocational training, and related infrastructure.

Furthermore, it is also arguable whether the expected high level of growth dividend is a realistic and sustainable option. Compared to China foreign direct investment has been small in India. India's investment environment still remains questionable in the minds of many. Bureaucratic delays, inefficiency, and poor governance practices are obvious stumbling blocks to investment. For example, even in Pakistan it takes only twenty-four days to register a company, but it is eighty-nine days in India.

India's image as one of the most corrupt countries in the world may also stand in the way of attracting investment (Volcker 2007). Besides, Lambsdorff's (2003: 229–43) analysis of the impact of corruption on net capital inflows found that a country's law-and-order tradition is a crucial factor in attracting capital. The examples of China, South Korea, and Suharto's Indonesia seem to support Lambsdorff's findings. The World Bank's (1997a: 103) argument that countries that have achieved high rates of economic growth despite serious corruption may find themselves paying a significant price in the long run cannot be denied, as has been the case in post-Suharto's Indonesia following the 1997 Asian financial crisis. Given India's poor reputation in this area and the difficult law-and-order situation in many parts of the country, one has to wonder whether foreign-investment-led growth can be sustained over time. On both counts—corruption and law and order—India remains vulnerable.

In such contexts questions remain whether a government can politically or financially afford to commit such huge investments on a project that has not been rigorously and transparently tested for financial, economic, and environmental viability with full participation of all stakeholders.

Is There an Alternative?

The alternative is to seek external funding through the multilateral funding agencies. Bilateral agencies would not be able to risk funding such a mammoth project, and private-sector funding is out of the question, as the expected cash flow would not be enough to service the capital repayments. But even if the multilateral agencies would consider funding the project, they would demand an open, transparent, and publicly verifiable EIS, and an assessment of the possibility of human rights violations that may be caused by the large-scale involuntary displacement (Cohen and Deng 1998: 62; Banerjee 2005: 307), and a host of other information that can be publicly verified. Such assessments would certainly explore alternative options available to achieve the same or similar outcomes.

This would require assessing the opportunity cost of the investment that would be required for the project. Assessing opportunity costs also forces governments to look for alternatives, as often such exercises reveal new ways to achieve a better outcome from investments. The Arun III hydropower project in Nepal is an example (Bell 1994: 113–15, and 1995, as cited in McCully 1996: 230; Lama 2001: 183–84).[17]

The opportunity costs of large investments also need to be considered in the context of other priority investment needs of the economy. Economists have identified eleven major problems facing the Indian economy (Parikh 2002: 8). Five of particular concerns are India's exports and the global trade regime under the WTO; the persistent fiscal imbalance; inadequate physical infrastructure, including a chronic power shortage; the inadequacy of social infrastructure; and the challenges of climate change.

Some other concerns are also relevant to such a large proposal. The global economic outlook is one. In the medium term, the global economic outlook remains confused. GDP growth rates in the two largest economies in the world have not been encouraging, though. In the United States the annual per capita growth rate during 1990–2002 was only 2 percent and in Japan it was only 1 percent. The United States has been running huge deficits in its external account, and these are showing no signs of that being brought under control.

The NGO movement also needs to be taken into account when evaluating the viability of the river-linking project. In spite of some major deficiencies, the movement is gaining strength everywhere. India has about 14,000 registered NGOs. They can greatly influence the way the international system tackles environmental problems (Blatter, Ingram and Doughman 2001: 8; Pawar et al. 2004: 14; and Wapner 2000). Given their experience from the Narmada movement, Indian NGOs will certainly ensure that all issues are openly debated, and there will be no dearth of popular support for their cause.

Like any large project, in the final analysis political support is the critical factor in the feasibility of the river-linking project. And when it comes to politics, NGOs are essential for mobilizing the community against proposals that could go against larger social interests. NGOs represent the people at the grassroots level, and there are many examples—from Asia to Africa—of people at the grassroots devising ways of managing renewable natural resources properly (Swaminathan 1986).

The most disconcerting issue, however, remains that, first, the World Bank found that both in terms of performance and economic viability considerations the outcome was poor for most of the pre-1990 projects (World Bank 1991: 9–10). Second, total installed hydel-power capacity currently stands at 25,100 MW, but actual production is only 74.5 billion KW hours. Given these miserable performance records, one is entitled to ask how realistic are the proposed objectives, and how confident one can be that even a fraction of these two major objectives will be realized!

In view of the above, an assessment of the true benefits and costs of the river-linking project is vital before one can proceed, assuming that the project is environmentally, socially, and politically acceptable (Mitra 2003). All true benefits and costs need to be identified transparently. In a rapidly changing global economic and political environment, transparency and improved governance practices have been emphasized as necessary by all development agencies to ensure sustained economic growth. Last but not least, it is almost certain that the project will crowd out other public investments and reduce the government's capacity to adjust expenditures in response to external developments, as was the case with Sri Lanka's Mahaweli dam project (Athukorala and Jayasuriya 1994: 79–81).

Notes

1. Assuming an exchange rate of $1 to Rs.45.00

2. In 2005 the inflation rate fluctuated between 4 and 5 percentage points. However, in 2007 it showed an increasing trend. Also, some commentators have questioned the government's inflation figures. In a scathing editorial, the highly respected *Statesman* argued that the GoI financial mandarins and politicians "would like the Indian public to believe average inflation in India is in the region of 5 per cent. . . . What seems certain though is that [the Finance Minister] may need to send his economists back for a refresher course, or at least make them buy some textbooks of monetary economics published in the last 25 years." *Statesman Weekly* (2006), July 1. Even the government's 2006–2007 Economic Survey raised concern in this regard. See *Statesman* (2007), "Healthy growth, but price spiral is a concern: Economic Survey"; February 28.

3. Most rehabilitation projects, normally a part of dam construction costs, failed, either because of poor design, poor execution, or simply corruption. Besides rising re-

habilitation costs, an area equivalent to between 5 and 13 percent of newly irrigated land is typically lost to reservoirs, canals, and drainage infrastructure in India leading to a rise in per-unit cost of irrigated land. Globally, cost per-unit of new schemes is rising by about two to three times more than the earlier schemes (see chapter 4).

4. XI Finance Commission Report: Chapter XIII.

5. Sen (2004), while discussing Tagore's (1913 Nobel laureate in literature) interest in childhood education argued "Tagore would not be consoled by the extraordinary expansion of university education, in which India sends to its universities six times as many people per unit of population as does China. Rather, he would be stunned that, in contrast to East and Southeast Asia, including China, half the adult population and two-thirds of Indian women remain unable to read or write" in (Hallengren 2004: 205). Also, see World Bank Report "Reaching Out to the Child" (2004: Chapter 6—Underwriting the Child's Development—Public Spending on the Child); Mustard (2002: 23–61); and van der Gaag (2002: 63–78).

6. *Registrar General of India*: See www.censusindia.net/t_00_006.html (February 15, 2005).

7. *Statesman* (2007), "Data belies health hopes," February 22.

8. *Statesman* (2007), "NHRC Demand"; February 7.

9. The combined gross fiscal deficit of the state governments in 2004–2005 (revised estimate) was IR 1,236,350 million, 4.0 percent of GDP. See http://rbidocs.rbi.org.in/rdocs/Publications/PDFs (May 2, 2006).

10. XI Finance Commission Report: Chapter XIII.

11. See "Mendacity, the govt. budget-constraint"; "Public debt, govt. fantasy"; and "A down payment on the Taj Mahal"; "On money and Banking" *Statesman* February 3 and 22, 2006; March 1, 2006; and *Statesman Weekly* April 29, 2006.

12. XI Finance Commission Report; p. 111.

13. *Statesman* (2004), "Pension tension"; October 29.

14. *Statesman* (2007), "India shares jobless growth: UN," February 21; and "Jobs crunch in organised sector"; March 28.

15. The report is titled "Asian Experience on Growth, Employment, and Poverty" and is prepared by Dr. A. R. Khan. Published in Colombo and Geneva: UNDP Regional Centre and International Labour Office.

16. *Statesman* (2007), "Job scenario grim," February 28.

17. Nepal's success in developing small hydro-electric projects against the wishes of the multilateral agency was made possible by the grassroots commitment of the government advisors and the support provided by the government. See, J. Emmons (1990: 90–92).

CHAPTER EIGHT

Political and Governmental Perspectives

Water and the Indian Constitution

The UN guidelines on managing scarce water resources stress that water policy should contain three critical aspects: legal, institutional, and managerial. It also suggests a number of additional guidelines. While a few of these guidelines can be challenged, they nevertheless provide a template against which each country can develop its own policies on water harvesting, processing, and distribution (UNWWDR 2003: 369–84). Of the three critical aspects, legal and institutional arrangements are usually set in countries' constitutional provisions. Detailed policies on water harvesting, processing, distribution and alike normally remain responsibility of the government of the day.

The Indian Constitution contains three lists: the Union list (List I), the State list (List II), and the Concurrent list (List III). Water has a place in each list. The lists are designed to identify the responsibilities of each level of government including those concerning water.

Initially the Constitution provided all water-related responsibilities to the Central and state governments only. The 73rd and 74th Constitutional Amendments in 1992 brought the local governments into the water-management arena to allow grassroots participation and accountability. Three key issue, however, have not been addressed: the specific responsibilities, relationships, and roles of the third-tier government with respect to the other two tiers of government and financial arrangements (Rajaraman et al. 1996: 1071–83). Even the National Commission for Reviewing the Working of the Constitution (NCRWC) remained silent on these matters.

The Union's responsibilities on water management include managing in the public interest the inter-state rivers and river valleys specified by legislation. The states' responsibilities include water supplies, irrigation and canals, drainage and embankments, water storage, and water power, subject to the provisions covered in the Union's responsibilities. Article 262 of the constitution with respect to water usage and management issues stipulates that:

- Parliament may by law provide for the adjudication of any dispute or complaint with respect to the usage, distribution, or control of waters of inter-state rivers or river valleys.
- Irrespective of other provisions and articles in the constitution, Parliament may by law provide that neither the Supreme Court nor any other court shall exercise jurisdiction with respect to any such dispute or complaint that might arise following the adjudication as referred to above.

The Constitution allots to the states the principal responsibility for managing internal water resources, although in reality the management of groundwater and inter-state rivers requires central government action. The central government has the constitutional responsibility to arbitrate when there is a dispute between one or more states in sharing the waters of common rivers, should they fail to come to a mutually agreeable arrangement.

Central clearance is compulsory for Concurrent list projects that do not fall explicitly into the "state domain." In reality, central clearance is required for all major and minor irrigation, hydropower, flood control, and multipurpose projects before these can be included in the national plan. Projects of significant dimensions also must obtain clearance under two specific central acts: the Forest Conservation Act and the Environment Protection Act.

Water Administration (Central Government Level)

The principal responsibility for freshwater management at the central government level lies with the Ministry of Water Resources. Its responsibilities include development, conservation, and management of water as a national resource, including overseeing the regulation and development of inter-state rivers. There are four other ministries with specific water-related responsibilities: the Ministry of Urban Development, the Department of Drinking Water under the Ministry of Rural Development, the Ministry of Power, and the Ministry of Environment and Forests.[1]

By political and operational necessity, four other ministries are also required to play important parts in water-related policies. These are the Departments of the Prime Minister, Home Affairs, Foreign Affairs, and Finance. In terms of political clout, they are four of the five most powerful ministries, the other being the Defence.

The prime minister is involved mostly as the head of government. The Departments of Home Affairs, Foreign Affairs, and Finance cover areas that specifically fall within their domains of policy expertise. For example, both the Department of the Prime Minister and the Home Ministry are involved in matters relating to inter-state relations, including inter-state disputes on water sharing. Similarly, international water disputes cannot be addressed without the expertise of the Foreign Ministry. And the Finance Ministry has responsibility for overseeing all expenditure and investment decisions.

In the central government, the place of the Water Ministry has varied over the years. The ministry has not always been within the Cabinet, and has experienced frequent changes of personalities that have often contributed to poor policies. On many occasions the prime minister was required to take a leading role in addressing inter-state water disputes.

The Water Ministry administers seventeen agencies. The organizational structure of the ministry, to put it mildly, is a bureaucratic quagmire. The responsibilities of the various parts of the agencies remain unclear. Even the functional relationship between various related agencies is not based on any sound principle. And the hierarchical bureaucratic structure does not suit the management needs of a sector that covers everyone in the community, rich or poor, urban or rural, sophisticated or traditional (Batley and Larbi 2004: 229–33).

The Parliament passed the Inter-State Water Disputes Act in 1956 under Article 262. In the initial years some inter-state disputes concerning water-sharing arrangements were settled relatively easily. Since the 1960s inter-state disputes have increasingly been assuming unmanageable proportions, however. There are many reasons for this:

- With increasing populations, states have been demanding a larger share of water from inter-state rivers, thus upsetting the existing share balance. This demand for a larger share has put severe strains on existing water-sharing arrangements. For example, in 1966 when the state of Haryana was created by bifurcating Punjab, Haryana demanded 4.8 million acre feet (MAF) out of the total allocated to Punjab, and Punjab, in turn, claimed total control of all the water (Corell and Swain 1995: 136);
- Since the constitution was promulgated many new states have come into existence and the demand for separate states is increasing. With

this, the number of riparian states increased, as did the possibility of inter-state conflicts over water sharing;
- During the first two decades of independence, the Union and the state governments were virtually led by Congress. All along Congress has been virtually managed by its High Command; consequently, it was able to keep a lid on many inter-state disputes;
- Congress no longer has undisputed control over the central or state governments, as during the last quarter of the twentieth century the monolithic Congress party, founded in 1885, started losing its somewhat hypnotic control over the Indian masses (Omvedt 1993: 177; Weiner 1987a: 42–43; 69–75). Coalition governments have their own imperatives to use the water issue for narrow political gains;
- Coalition politics are more prone to using water as bargaining chips in their power-sharing arrangements. As a consequence politically difficult issues are often left unaddressed, resulting in simmering discontents that contribute to the hardening of the attitudes of disgruntled parties; and
- A strong and visionary political leadership could have contributed to resolving many such disputes, but none exists. This minimizes the chances of an amicable resolution of disputes over water-sharing arrangements.

Inter-state Water Disputes

Since independence there have been a number of inter-state water disputes. The one most discussed is between Tamil Nadu and Karnataka over the Cauvery waters. This dispute has now been going on for almost fifty years. It has flared up intermittently and often taken ugly and violent turns (Iyer 2003: 49). The second most prominent dispute is between Punjab, Haryana, Himachal Pradesh, and Rajasthan on sharing the Ravi-Yamuna waters.

Unprincipled political leadership, its need for survival, and (sometimes) coalition governments have kept the Cauvery dispute alive. The dispute generated so much bad blood between these two communities that a member of the Indian Parliament has allegedly threatened that if the Cauvery water-sharing dispute between Tamil Nadu and Karnataka is not resolved, Tamil Nadu could go the way of Kashmir.[2] Nearly two decades after its formation, the Cauvery Water Disputes Tribunal gave its final verdict in February 2007, and lion's share of the water was allocated to Tamil Nadu.[3] This promptly led to political uproar in Karnataka and a countermovement in Tamil Nadu. The Speaker of the Lower House of the Indian Parliament convened a meeting of the members from the disputing states to decide if Parliament could discuss the water row. Unfortunately, the meeting ended abruptly in the face of the anger and frayed tempers displayed by members from both states.[4]

Worst is the position taken by a former prime minister (who comes from one of the disputing states) when the Speaker of the House allowed him to intervene on the "Motion of Thanks" to the President's address on the tacit understanding that he would not raise the "emotive Cauvery-water-sharing" issue, an understanding he did not keep. This led to a heated discussion and protests led by the Members from the other state that forced the Speaker to politely admonish the former prime minister and allowed discussion on this issue after completion of all financial business before the House. This also forced the Prime Minister to appeal to all political parties to treat water as a national resource critical for sustained development in the years to come (Rajamani 2007). This incident clearly demonstrates how difficult it is to handle the inter-state water disputes without invoking raw emotion, which could violate all Protocol.

The Ravi-Beas-Yamuna water-sharing dispute has also been continuing for a number of decades, and has taken a turn for the worse in recent years. The Punjab Reorganization Act 1966 and the Eradi Tribunal, which was set up in 1985 following the Rajib Gandhi and Sant Longowal Accord of 1985 (Iyer 2004), allocated water among the affected states. However, the Punjab Government passed the Punjab Termination of Agreement Act 2004 unilaterally. The President referred the act to the Supreme Court under Article 143 of the constitution and the Court rejected the Punjab position. This unilateral action by Punjab raises the question of India's nationhood, with one part of the federation unilaterally revoking an agreement that was the product of an international agreement between two sovereign states (the Indus Water Sharing Agreement of 1960); and an agreement between the central government and the state. Punjab's actions would not have reached the current level if evaluations had been carried out on Punjab's long-term land capability, environmentally and economically. This highlights the fundamental need to consider water and land use together, and to bring efficiency and sustainability to the forefront of the water-policy debate.

While states want more water to increase their irrigation capacity to produce more food for their growing populations, they remain or choose to remain unaware that marginal productivity of canal-irrigated lands has been declining over the years in many parts of India (as found by Dhawan [1995] and Rao [1985], cited in Moench [2002: 149]). By exaggerating the potential area under irrigation by including unsuitable lands, the real cost of output has often been grossly understated (Ram 1994).

States have the constitutional right to claim more water, but their claims must be based on both state and the national interest, and claims must be objectively justifiable. Punjab is already suffering from a high incidence of waterlogging and land salinity. This depresses agricultural yields and makes the land sterile, with enormous consequences for the local and national economy and public health in the medium-to-long run. Before demanding additional water, Punjab should have seriously considered whether it would require additional water for irrigation, given its land conditions and the environment.[5]

There are also other water-related disputes, many of which could become major if not handled judiciously and expeditiously. One such dispute looming in the horizon is between the states of Andhra Pradesh and Maharastra relating to the inter-state Babhli irrigation project.[6] Disputes will also continue to arise, as new states are created. The absence of political leaders of sufficient national stature who could encourage conflict resolution through the political process is also contributing to the continuation of disputes (Shastri 2001: 279–89; Arora 2004: 171–73). These disputes will not go away, given that during the last three decades there has been a growing tendency for many states to take a very narrow approach on wider issues of national interest in a way that was unthinkable earlier. This was not entirely unexpected, as widening poverty results in disentitlement, deprivation, and disempowerment. Growing coalition instabilities and compromises only encourage concentration on narrow political gains at the expense of national interests (Yasin and Sengupta 2004: 2).

The ferocity of many of these conflicts would have been much less if *water related* issues were considered in a more objective basis. There are four basic steps essential to integrated water management and the conservation of water resources: commit, understand, evaluate, and do (Paul 1989: 193–97). For Punjab, Haryana, and Rajasthan, in particular, good water resource management is not necessarily the same as in temperate or tropical countries. These four steps would have enabled them to develop water policies to prevent desertification, waterlogging, and salinization (Falkenmark, da Cunha, and David 1987: 94–101).

Primary salinity in land develops naturally, mainly in areas where the rainfall is insufficient to leach salt from the soil and evaporation is high. Excessive irrigation makes the situation worse; the Australian experience and that of the Colorado basin in the United States are examples (Murphy 1999; State Salinity Council 2000: 14–15). So are those of the states of Punjab, Haryana, and Rajasthan. Dryland salinity also has enormous negative economic and financial effects on physical infrastructure, such as the road and rail networks (Short and McConnell 2001: 32–33 and 44–46).

The British introduced canal irrigation in Punjab, which has done huge damage to the soil's productive capacity. Satellite imagery provides a warning that must be heeded before there is further intensive cultivation with irrigation water if future disaster is to be avoided. The Ganges plains in Haryana and eastern UP are already showing symptoms of such disaster (Chapman 2000: 109; Chapman 1992: 30–31; Bradnock 1992: 61–62; Wade 1988: 222).

Nationally, there are another eight major unresolved inter-state disputes. Some states involved in them have tended to exhibit extreme selfishness and an unwillingness to look at the issues scientifically and from a wider national perspective. The reasons for such attitudes appear to include the following:

- The freshwater demands by states are always for more water than is available;
- Within each political jurisdiction, demand is increasing with increased levels of economic activity, including demands for irrigation water and a higher standard of living. States with a dominating agricultural sector tend to make unrealistic claims for the available water and are scarcely willing to consider alternatives. In the process, water becomes a political issue, and political parties are loath to go against the wishes of their electorate even when those wishes are questionable or unjustifiable;
- Policy makers generally have failed to articulate the relevant issues or to educate stakeholders on the importance of treating water as a scarce natural resource that needs to be considered together with environmental, financial, and sustainability concerns; and
- Policy makers have also failed to keep stakeholders informed of rapid changes that are taking place in global freshwater management policies. It is arguable whether policy makers themselves are aware of such changes. Their unshakable faith in supply-side policy alone suggests that they are not. This has created a convoluted psychology whereby consumers have become accustomed to inefficient water use and management practices that cannot easily be changed by management alone, and whereby vested interests take advantage of such situations and force unscrupulous politicians to follow the line.

The Inter-state Water Dispute (ISWD) Act of 1956 worked well in the early years. Four phenomena helped to resolve disputes:

- The euphoria of independence still dominated the political psyche of the Indian population at large;
- Until the election of the first non-Congress government in Kerala in 1957, Congress or its political ally held the government in all states and

at the center. This made it easy to tackle complex problems at both the political and administrative levels simultaneously;
- The political dominance of Congress allowed government to persuade warring states to accept centrally bargained decisions, even though they might not have fully satisfied them. With the rise of regional parties and the growth of coalition governments, such opportunities have all but evaporated; and
- Although the quality of the Indian body politic during the first few decades after independence was not the highest, there was nevertheless an overall commitment to preserve the sanctity of the constitution and the nationhood and the basic tenets of a pluralist democratic society. This has been lacking in recent decades.

Following the decline of Congress's fortunes, governments in many states have been formed by political parties of different persuasions than the party at the center. Consequently the central government has increasingly become captive to the demands of coalition politics, resulting in political parties with few members wielding power disproportionate to their power base. An analysis of the election outcomes since 1989 confirms this trend.[7]

Furthermore, coalition politics is institutionally conditioned and governed by anticipation; no act of coalition politics can be understood in isolation from other acts, and the rules vary from decade to decade and party to party. Also, the more parties there are in Parliament (other things being equal), the more complex the coalition-bargaining environment (Müller and Strom 2000: 1–33, 560). The Indian Parliamentary system is based on the British model, which is generally considered a source of *adversary politics* that often stands in the way of the development of a coherent policy base, particularly when the preconditions essential for a flourishing pluralist democracy are missing, and this is not helping the situation either (Colomer 2001: 209; Shonefield 1965: 99–102, 402; Pemberton 2004: 192; Balfour 1936).

There are also entrenched caste considerations and a social system structured by three sets of asymmetrical relations in India (Mukherji 1986: 50). These factors have created contradictions within various coalitions and with the rising aspirations and expectations of the people, generating many conflicts, some of which have also assumed ethnic dimensions (Phadnis 1990: 256). Some argue that this is inevitable, as the Indian political system that emerged after freedom was not dominated by true representatives of the people but by an indigenous capitalist class (Bhambri 1986; Varma 1986). Other studies lend credence to this view (Mollinga, Doraiswamy, and Engbersen 2004). National water policy is, of course, formed in this environment.

Besides inter-state conflicts, there are other types of water-related conflicts, such as conflicts between municipal needs and agricultural demands, conflicts between agriculture and industry, and those between industry and the community. These conflicts escalate with the reduced availability of freshwater.

Appeal to National Integration: Will It Work?

As India moves from being a water-stressed country to a water-scarce one, conflicts between states will increase. In such an environment, calls for *national integration interest* may sound hollow. India's internal conflicts during the last few decades, whether related to water sharing, land tilling rights, movement of mineral resources, or job preservation, are not showing any signs of waning. And while the ethnic structure of politics at the center has remained diffuse, at the state levels ethnic politics has assumed sharper dimensions. In some states the demands of the dominant ethnic community have led to direct confrontations with the center (Phadnis 1990: 254). Some of these states remain parties in ongoing water conflicts. Worse, it has been found that even at the village level people do not hesitate to divert water through their own field channel network, although they may not need the water, to ensure that the people in the next village cannot exert pressure on the irrigation department for more water (Wade 1988: 221).

As discussed in chapter 4, India's water need is dominant in the irrigation sector. Effective government policy in that sector requires the existence of strong institutions which, in turn, require strong and unshakable political commitment backed by penetrating research and a committed administration. Scholars have been emphasizing the need for institutional reform, without which water and agricultural productivity cannot be managed effectively (Mitra 1996: A31–A37; Dinar, Balkrishnan, and Wambia 1998: 4–22). But they have done so with little success.

Three factors in the inadequate reform of the irrigation and agricultural sectors are paucity of resources, poor performance of existing major and medium-sized irrigation systems, and entrenched political and bureaucratic forces within institutions (Paul 1990: 29). Among many, failure to implement effective land reform throughout the country (World Bank 1997: 12) remains a stumbling block in achieving substantial reform and inclusive growth in the agricultural sector (Stiglitz 2007), which consumes about 90 percent of the available fresh-water. In highly politicized developing countries such as India or the Philippines (Frederiksen, Berkoff, and Barber 1993: xix; Kerkvliet 1990: 11) effective reforms in this sector are not easy, to say the least.

It is clear from the fact that the dispute-resolution recommendations of the Sarkaria Commission, which was established by the government to review center-state relations, have not yet been fully implemented. However, the recently promulgated Inter-State Water Disputes Act, which makes tribunal decisions legally unappealable, is an improvement over past practices. Still, the basic issue remains that water sharing is as much a matter of law as of environment, land use, and other socioeconomic considerations. It is, therefore, preferable that all inter-state matters, particularly water-sharing issues, in dispute be treated harmoniously within one national institutional-framework instead of by separate tribunals, which run the risk of pursuing inconsistent approaches.

The main administrative thrust therefore has to be initiating changes which will increase policy makers' capacity to solve both current and future challenges. This will require a clear understanding of the operational linkages between the various components of water institutions, including water law, water policy, and water administration, and the capacity to mobilize maximum political acceptability of the policy outcomes (Saleth and Dinar 2004: 302–28).

The establishment of NCRWC was an important initiative. It recommended, among other things, that the River Boards Act of 1956 be repealed and replaced by a comprehensive enactment. It was a good recommendation, as the river boards that were to be established under the 1956 act would all have operated in an advisory capacity without any teeth, because states affected by adverse recommendations would have resisted them on the grounds of protecting states' rights.

This was the implicit reason for the central government's inability to establish such boards earlier, although it had the power to do so. The NCRWC's view that inter-state rivers are material resources of the community and national assets, and by implication that the national interest should determine how waters of such rivers be allocated, is indeed a bold and sensible approach. As this is a fundamental shift from the original constitutional provision, a more participatory consideration of it by all segments of the community would have made it easier for both the Commission and the government to specifically determine the mechanisms for its implementation.

NGOs are usually successful initiators of social change provided they remain actively focused on people's concerns and needs (Shankar 2004).[8] They have been able to make an increasing number of people agents of change by insisting that state policies focus on people (Breitmeier and Rittberger 2000; Gan 2000). NCRWC could have used NGOs to educate stakeholders and in the process made it easy for the government and others to initiate the much-needed change in outlook. That would have saved the criticism that its recommendations had no practical value since it failed to suggest mechanisms to reconcile states' rights with the national interest (Iyer 2003: 73).

In the absence of a structured national law, each state can go its own way to develop laws that are applicable within its boundaries. Individual state laws could create more problems than they solve, as complex interrelationships such as those needed for pollution control, water treatment, the use and discharge of affluents, and the monitoring and enforcement of standards require one comprehensive legal base. Such situations can pose a threat to national integrity. The need for national water laws was emphasized by all participants in a series of national seminars organized by the National Commission for Integrated Water Resource Development (NCIWRD) in 1999. Even this has not been incorporated into the national water policy.

A Tricky Constitutional Issue

India has maintained its democratic system of government since independence, except a short period during 1970s. Regular elections have been held at all levels. However, the institution of the judiciary in India has had mixed success over the years. On occasion its judicial activism has been questioned by constitutional purists. The failure of other institutions to perform their constitutionally assigned responsibilities properly has in most instances contributed to its judicial activism (Kashyap 2004: 31). This activism became increasingly pronounced in the 1980s when it started hearing applications under public interest litigation (PIL).

The Court's activist position is based on defending constitutional integrity, democratic institutions, and democratic practices (UNDP 2002: 172). The Court is aware of the pitfalls and dangers of taking this path, as is clear from a former chief justice's caution to the legal community against misuse of PIL (Ray 1999: 63). Another former chief justice also cautioned against judicial creativity in the form of judicial activism, but stressed that such activism is an essential aspect of a constitutional court. He argued that when the executive branch refuses to apply law wilfully and conspicuously refuses to do its duty, it falls to the judiciary to act in defense of the constitution, the rule of law, and equality before the law.[9]

However, the legislative branch (the lower of the Indian Parliament) in recent days has described the Court's performance as an impingement on Parliament's sovereignty by *judicial over-activism*.[10] It is possible to argue that if both the executive and the legislative branches were performing their constitutionally assigned responsibilities in line with the spirit of the provisions of the constitution then possibly there would have been no need for many of the Court-decisions, which reflect some degree of *judicial over-activism*. The Court's pronouncement on the river-linking project undoubtedly falls in this category. The tragedy, however, is that in the context of this project the roles of both the executive and legislative branches remain highly questionable. There is no sign yet on the horizon that either of these branches is firmly committed to examine all project-related issues transparently before proceeding further with this project.

The Supreme Court and the President are now entangled in controversy over this issue. On a PIL application the Court passed an order to the GoI to complete all the proposed links within a period of twelve years and sought periodic reports on progress of the work. Though all three tiers of government are constitutionally responsible for water management policies, it is questionable whether the central government's Court-mandated role is constitutionally enforceable. The GoI responded to the Court that it was trying to develop a consensus between all the state governments and was waiting for the final feasibility report on the project.

There is a danger that the feasibility report will be extremely sketchy and may not cover all issues. Given past experience, the GoI could well use the excuse that since it was required to respond to the Court order, time did not permit it to consider all issues relevant to the river-linking project. And the Court could issue further directives to implement the project on the basis of the GoI's response. But what about the 72nd and 73rd amendments of the constitution? They give equal say to local governments on water management policy.

Another important issue is the basis of the Court's decision to set a twelve-year timetable to complete the project. How did the court arrive at that timetable? It has neither the capacity nor the expertise to make a unilateral decision on such a complex issue. A former government advisor and activist has questioned the Court's position (Iyer 2002a: 4595–56).

The President has shown a keen interest in the project and has sought details from the government.[11] A scientist of his stature is naturally interested in such an engineering marvel, although constitutionally it makes for an interesting question: is it proper for him to get directly involved in an area where the elected government should have the final say? He should also be aware that the project involves more contentious matters than meet the eye and that these need to be resolved.[12]

Finally, one has to ask whether one or even a few persons, unaided, can intelligently make this kind of judgment. To be a useful guide to action, such judgments must pool the ideas of many people (Elkin 1996: 202).

Vested interests could use this revocation to threaten India's nationhood by spewing venom similar to that of Rahamat Ali (1942: 3) against Indianism while championing the cause of Muslims. The bloody conflict for the creation of a separate Khalistan not too long ago was another example. The wounds of 1984 from the Khalistan conflict will not go away easily, particularly in the current socioeconomic political environment (Breman 1999). The only way to tackle such atrocious attitudes is with intellectual honesty and constant vigilance against all forms of injustice (Marden 2004: 266).

Notes

1. Often new departments are created and also departments' nomenclatures are changed.

2. *Economist* (August 24, 2000), "Water in India: Nor any drop to drink": "Their (states) quarrels are bitter. The southern state of Tamil Nadu claims that neighbouring Karnataka violates commitments to share water from the Cauvery. 'If the Cauvery problem is not solved,' wrote one Tamil Nadu MP this month, 'Tamil Nadu could go the way of Kashmir'"; pp. 49–50.

3. *Statesman Weekly* (2007), "Cauvery: TN gets lion's share": The Cauvery Tribunal allocated 74.6 percent of Tamil Nadu's demand and 58.1 percent of Karnataka's demand. Following the pronouncement of the verdict political maneuvring by the states has risen to a new height; February 10.

4. "Bangalore road closed to TN," "Gowda reservations over Cauvery award," "Karnataka ready for Bandh," "Cauvery cauldron set to boil for TN as well," and "Centre in a spot over Cauvery" (*Statesman* 2007: February 9, 12, 12, 15, and 24 respectively); "No headway in Cauvery dispute" (*Statesman* March 7, 2007).

5. British Broadcasting Corporation (2007), "Feeding the World"; the BBC World Service series investigates the growing but often underreported challenges facing the world's food supply. The reporter went to Punjab's rural heartland and found that many farmers were in a desperate situation because of rapidly declining water level and land-salinity (Growing Pains in Part One); http://news.bbc.co.uk/2/hi/programmes/documentary_archive/6500041.stm (March 29, 2007).

6. *Statesman*, "Will YSR stand restart a water war?"; http://www.thestatesman.net/page.news.php?clid=2&theme=&usrsess=1&id=151274 (March 29, 2007).

7. India follows the British system of election: first past the post gets elected. It does not ensure that the winner really reflects the community view, as often with so many parties and candidates contesting election winners are elected with a handful of minority votes. For example, even when Congress virtually won a landslide victory in 1980 by winning 351 seats, it only secured 42.7 percent of the popular votes (Weiner 1989: 216). In the current Parliament (Lower House) the ruling Congress (145 seats) alliance (with fifteen other parties) and the opposition Bharatiya Janata Party (BJP: 138 seats) alliance (with nine other parties) together constituted twenty-six of the

forty-one political parties that won seats in the election. Two Communist parties (Congress's main ally) have fifty-three seats. Each of the remaining thirteen Congress-allies on average won only 1.7 seats, and the BJP allies won on average 5.2 seats. Large numbers of political parties and independents contesting elections thus made it possible to get elected by securing a small number of votes. Also, see Weiner (1989: 87–950; and Chandra 2004: 87–125). These, in the long run, often contribute to a poor policy development environment.

8. *Statesman* (2004), NGOs need to be *people* oriented, not just sub-contractors of *donor* or *funding* agencies, July 10.

9. *Statesman* (2005), "CJI cautions against 'pro-active role' of judiciary"; January 24.

10. Statesman (2006), "MPs to debate *judicial activism*," December 8.

11. *Financial Daily* (2003), "Interlinking of rivers to be driver of growth: Kalam," Chennai, February 23.

12. NIRAPAD (2003), Newsletter Issue 6; Dhaka.

CHAPTER NINE

Regional Perspectives
Bangladesh, Nepal, Pakistan, and China

A little more than one-third of India's freshwater flows from outside its national boundaries to the north consists of both rain- and snow-fed water. The impact of climate change on them remains unknown at this stage. But the recession of the Himalayan glaciers and insufficient snow accumulation, in particular, are likely to cause these flows to decline.

To be effective, India's freshwater policy also needs to take cognizance of the policies of the other riparian countries in the region. The 1997 UN Protocol on sharing riparian waters has not yet been ratified by the required number of countries, but it can't be entirely ignored because global public opinion broadly accepts Lord Macnaghten's (1893) view that water does not exclusively belong to one legal physical and political entity. A regional approach incorporating river basins can contribute in ensuring an optimal solution for all affected countries within the context of the 1997 draft UN Protocol, as water management closely relates to land, population, and trade policies. It is unfortunate that the mainland sub-continental countries have never been able to develop an ethos that put their common good above their individual interests (James 1994: 61–69).

India has the largest share of the Ganges basin, followed by Nepal and Bangladesh. Both these countries are among the poorest nations on earth and their economic interests require that the waters of this river are used optimally. Nepal's main interests are to minimize the threat of flooding during the monsoon season and to use water for power generation and irrigation. Bangladesh's needs are to deal with annual floods and to ensure irrigation water during the dry months. It cannot escape that to deal with the flooding situation successfully its efforts must involve integrating development programs with that of the upper riparian countries.

During dry months, water flows in the Ganges fall significantly. The densely populated Gangetic basin in the north and eastern India are increasingly demanding more water from the system, causing this drop in flows in the lower basin when the river reaches Farakka before entering Bangladesh.

Low water flows affect both India and Bangladesh, but they are ambivalent about how to augment the flow in the Ganges. They agree, however, that some of the Brahmaputra River's surplus water may be used. As nearly one-half of the Brahmaputra's catchment area belongs to China, its actions on the upper stream of the river would be critical should these two countries decide to divert the Brahmaputra's waters.

Pakistan, whose main source of freshwater is the Indus river system, has the largest share of the Indus basin with about 57 percent, followed by India with 26 percent and China with about 11 percent. The 1960 Indus water-sharing treaty brokered by the World Bank has worked well so far, but in recent decades problems have emerged, as both India and Pakistan are facing increased demands for freshwater. Pakistan has a major problem on its hands already, as when the Indus reaches southern Pakistan it hardly carries any freshwater. A significant section of southern Pakistanis believe that Pakistan's own freshwater policy has contributed to this problem, as it allows harvesting more water from the system by the powerful Northern Province of Punjab.

India plans to harvest more water by building dams on the Indus tributaries. Pakistan opposes this on the grounds that it would adversely impact its water flows. India's Baglihar proposal on the Chenab was sent to arbitration, per the treaty, and the dispute has been resolved.[1] Pakistan is also planning to harvest more water from the Indus system by building a new dam (Bhasha dam) in Gilgit in Pakistan-controlled Kashmir, which India opposes.

China owns a little more than 11 percent of the Indus basin but has not made any claims or used its waters yet. This remains a dormant issue at the moment. China is facing a severe water shortage, estimated to be 40 BCM a year. About 400 of its 669 cities are already facing water shortages and some 20 MH of its farmland are affected by drought.[2] In light of the 1997 draft UN Protocol on using waters of riparian rivers, a claim by China in future could not be dismissed outright. Should China make a claim, the issue would need to be addressed at the basin or regional level.

Like the Indus, the Brahmaputra River comes out of Tibet and flows through India. Nearly half of its basin belongs to China, 31 percent to India.

Both Bhutan and Bangladesh also own a small portion of the basin. Unlike the Ganges and Indus waters, the Brahmaputra's water has mostly remained unused, but demands for it are high. Should Bangladesh and India act bilaterally by diverting parts of the Brahmaputra's water to the Ganges system during the dry season, China could thwart them by intercepting the Brahmaputra's flows in the upper stretches of the river, although that would go against the spirit of the 1997 draft UN Protocol. Solutions to such a tricky issue can only be devised by addressing the issue at a regional level involving all the riparian countries.

Bhutan, a poor, land-locked country, is mostly covered by forests, and the northern part is under permanent snow. Some of this snow feeds the Brahmaputra. The recession of northern glaciers because of global warming is causing two problems there: a decline in the water flows to the river and exposure of the land to erosion. Bhutan's ecosystem is sensitive, and large-scale deforestation in the upper reaches and the retreating glaciers are causing massive erosion and devastating floods in India and Bangladesh (Mukherjee 2001).

Freshwater Resources and Political Complexities in South Asia

Globally, there are now close to 200 resource-related conflicts (Strauss 2002: 11). Most are over control of the resource. More than one Middle Eastern war has been fought over water (Plaut 2000: 19–20). Unfortunately, water-related conflicts in the sub-continent have the potential to become every bit as nasty as those in the Middle East.

Besides water flow through the riparian rivers, another critical source of conflict is the contamination of water by neglect or accident (Ward 1997: 67–68; Austin 2004: 272). Given the poor management practices of the sub-continental countries and the absence of proper regulatory mechanisms there (and in China), the possibility of such conflict remains extremely high. This conflict could, in turn, trigger other conflicts (Friedkin 1987). Unfortunately, there is no hard-and-fast rule on water quality because of differences in soil texture and other factors, and experts in this part of the world often disagree on technical information about water flows and quality (Manzur 1973: 251). Solutions to such problems need to be carefully drawn, as a new solution to an emerging problem can lead to more bitter controversies, which are often difficult for smaller nations with limited resources to foresee.

The 1947 partition of the sub-continent left deep scars on the psyche of the sub-continental populations and their leadership. These scars have been deepened by ingrained hostilities, communal tensions, infiltration through the long porous borders, and global power plays. Bangladesh, which came into existence as a secular country, is now an Islamic country and has distanced itself from the concept of a secular state (Catherwood 2003: 209). This and other porous-border related issues have created tension with India that has inhibited the development of mutual trust, which is essential to managing the freshwater policies of both countries. This is clearly reflected in Bangladesh's internal policies. A group of former Bangladesh freedom fighters have complained to the Indian journalists recently that the country's military had been thoroughly infiltrated by Islamists, especially of the Jamat-e-Islami variety who favoured that Bangladesh should have a loose confederation with Pakistan. They further claimed that to Jamat-e-Islamists *freedom fighter* is a dirty word.[3] India is an upper riparian country to most rivers in Bangladesh, and both real and imaginary conflicts exist between these two countries on water-sharing and related arrangements.

No major direct water-related conflict currently exists with Pakistan, but India has fought three wars with Pakistan over Kashmir. The Kashmir issue has influenced Pakistan's policy approach in resolving other contested issues with India, notwithstanding the enormous global pressure to resolve bilateral disputes either through negotiations or arbitration. Both countries remain steadfast on their respective positions and any future disputes concerning harvesting additional water from the Indus system has to be seen in this context.

Nepal and Bhutan do not have any conflicts with India similar to those of Bangladesh and Pakistan. Nepalese activists have openly complained that when India built a barrage on their soil to stop the Kosi River from flooding, the problem was simply shunted upstream, thereby becoming Nepal's problem. Its law that all treaties on the sharing of natural resources must be approved by the Nepal Parliament was aimed at India. Nepal is currently undergoing political upheaval, and political parties involved in this upheaval do not have a common approach on most economic, social, and political issues. Their contradictions may spill over to their approaches in addressing resource sharing arrangements and related issues with India. Past experience indicates that such possibilities remain high.

Shared River Basins:
Mutual Suspicion Dominates the Thinking Process

India must have water-sharing arrangements with Bangladesh, Bhutan, China, Nepal, and Pakistan. Yet the governmental systems of these countries are different than India's. The situation is further complicated by discrimination, exploitation, and oppression in at least four of the five countries. These asymmetrical relations are often used by governments for political advantage and often shape their approaches to contentious issues such as water. Most of the time they are a stumbling block to addressing water-sharing issues in a regional or river-basin context. Among other things, they have affected Bangladesh's approach to water-sharing arrangements (particularly with respect to Farakka) and Pakistan's dispute with India over India's proposal to build additional dams on the Indus system.

Smaller countries often tend to distrust their larger neighbours, and negotiations between a larger county and a smaller country inevitably favour the larger country (Hundley, Jr. 1966: 180–85; Dasgupta 2000; Ray 2004, 2004a: 57–80). China has been more successful in managing relations with smaller neighbours than has India.[4] On occasion India has allowed distrust to develop by virtue of its own deeds. For example, even though the feasibility study of the river-linking project has not been completed, the Supreme Court, the President, and the Prime Minister—in unison—expressed strong desire to complete the project soon, despite the fact that it would involve interfering with the flows of major rivers (Iyer 2002: 4595–96).

Any claim that India has yet to make a decision on the project and is awaiting the outcome of the feasibility study naturally would not be believed by her small neighbours. Such a claim provides opportunities for China, in particular, to court the support of these countries in its quest for regional political and economic supremacy. The possibility of this power play cannot be ignored.

The sub-continent is unique in the sense that these smaller countries are neighbours of two giant countries with different systems of governance who are keen to have their economic and political influence extended. Given Bangladesh's and Pakistan's historical suspicion of India's motives, and Nepal's uncertain political environment, a regional or basin-wide approach to water-sharing arrangements appears to be a distant possibility at this point, despite its unquestionable merit. No such approach is possible without China's direct participation.

It is difficult to predict how political relations between the regional countries will evolve over the coming years. Relations between India and China, and the future political and social directions of the two Islamic countries, Bangladesh and Pakistan, will be critical in shaping them. So will the outcome of the current power plays by the political parties in Nepal. The Bhutan situation is different from these countries, as the country has now established a *constitutional monarchical* system of governance and has updated its friendship treaty with India.

Not many parts of the world have such a complex geographical and political configuration as the sub-continent, except perhaps the Balkan countries. In this region there are states with extreme poverty; rival states that are achieving high economic growth rates; states with strong religious bases whose governments are often prone to taking decisions based on religious stances rather than liberal-pluralist principles; and states that are loath to bring about real change in the nature and institutions of globalization (Pacific Institute 2005). Furthermore, the spectre of climate change is now a reality and the world public opinion has targeted it as a priority area for urgent policy consideration. If nothing else, the forecast that inaction now may lead to an economic loss of between 5 and 20 percent of the global GDP each year (Stern 2006) must ring an alarm bell to all policy makers in the region, particularly to the poverty stricken sub-continental countries.

Notwithstanding differences of opinion between economists, it is absolutely necessary for the policy makers that they address water-sharing issues at a regional level, and failure to do so will be economically and environmentally disastrous for the entire population in the region in the near future (Zedillo 2007). The responsibility lies with both the big and small countries; prudence is key (Harries 2005: 15–17). In Appendix Three the sociopolitical environment in each of these countries is highlighted, as their policies are usually determined mainly by their internal needs.

Notes

1. *Statesman* (2007), "Experts clear Baglihar, for reduced dam height." The print media reported that both India and Pakistan have accepted the independent arbitrator's recommendations, and both have claimed victory. The report has recommended the reduction of the dam height by 1.5 meters; February 13.

2. These figures are collated from different sources. And, hence, these should be considered only indicative. However, they broadly tally with the figures available from Wikipedia: http://en.wikipedia.org/wiki/china_water_crisis (November 29, 2007). Also see J. McAlister (2005), *China's Water Crisis*. Deutche Bank China Expert Series, L55, Gheungkong Centre, 2, Queens Road Central, Hong Kong.

3. *Statesman* (2006), "*Conscious policy* to Islamise Bangla military" December 17.

4. *Statesman* (2007), "Gas pipeline" March 22.

PART THREE

THE WAY OUT

CHAPTER TEN

Reflections

India's freshwater policy is sketchy, internally inconsistent, and devoid of a long-term strategy to tackle the rapidly approaching water crisis. Its exclusive focus on supply-side solutions is unsustainable economically, politically, and, above all, environmentally. Policies adopted to date have not been accompanied by an improvement in the governance of water resources and water services (Briscoe and Malik 2006: xvi).

India's first national water policy was not formulated until 1987, when the country faced a serious economic crisis, though India had been investing heavily in multipurpose dams since independence. The dams provided short-term benefits but have had huge costs. Two prime ministers, Nehru and Rajiv Gandhi, publicly conceded that dams have not provided the expected benefits (McCully 1996: 168 and 179), but neither changed the direction of the policy. The second national water policy was formulated in 2002, but the fundamental focus did not change.

Both policies were flawed on many counts: they did not consider water as a scarce natural resource; they failed to address water issues in their totality; their premises were not evaluated; and stakeholders were not involved in policy development, though their role is vital in making policies work. Vested interests and the unscrupulous have taken full advantage of this situation.

The issue of sustainability with rapidly rising demand and global climate change has occupied center stage in the water policy debate everywhere. While the imbalance created by the rising demand could be foreseen, the impact of climate change on water supply and water use is less clear. But it could be severe. Consider the following:

- The Himalayan glaciers that feed the three major Indian rivers (the Ganges, Indus, and Brahmaputra) and a host of small rivers and streams have already receded considerably and are still receding;
- The expected rise in global temperature of 1.4 to 5.8 degrees Celsius by 2100 will heavily impact on evaporation and seepage rates, human health, agricultural productivity and national security. These are no longer long-term issues—they are already upon us; and
- Top soil, India's most precious resource (Wade 1985: 485), is being washed away at an alarming rate. The situation will deteriorate further with increased rainfall intensity, rising temperature, and receding glaciers.

As a little more than one-third of India's freshwater comes from outside the country, India's water policy cannot be developed in isolation of water policies of riparian countries. At a minimum, this will require a basin-wide integrated approach to water planning. Furthermore, other related factors, which need to be taken into account are: first, national standards and policies can no longer exercise absolute control over the national economy in the modern day and age; second, neither can they protect people and their environment, and finally, the state's role as an intermediary between the local and the global standards also appears to be diminishing (Ghai 1997: 25–45).

Given that many rivers in the Himalayan region crisscross sovereign nations, a bilateral approach may create unmanageable political and management problems in the long run. It is also worth remembering that water is an emotionally charged issue in the sub-continent, as about 90 percent of freshwater is used in the rural sector, where most of the population lives; rationality often becomes a casualty in the debate over it. Even national interests may not be enough to overcome this (Blatter, Ingram, and Doughman 2001: 10–11). Hence a regional approach is preferable, although rapidly emerging environmental concerns may require a biospheric approach in the not-too-distant future to accommodate such concerns.

In India's case regional considerations are particularly important because China, like India, suffers from a freshwater shortage and even a small increase in per capita demand of water there will be likely to adversely impact on the flow of water to the lower riparian countries, as two of the three Himalayan rivers originate in Tibet. China voted against the 1997 draft UN Protocol on managing freshwater of riparian rivers, and at least theoretically it remains outside the bounds of international regulations and protocols on sharing riparian-waters.

India has treated the rivers that originate in Bhutan and Nepal and which flow into it as individual entities and has dealt with Bhutan and Nepal bilaterally. This may not produce the desired outcome, as these rivers feed the larger rivers in the sub-continent which flow into or flow through a third country. Issues such as the quantity and quality of water, therefore, need to be considered in a wider regional context.

Bangladesh's and Pakistan's relations with India have remained frosty even in the best of times, and both are lower riparian countries to India. Bangladesh often finds it easy to blame India for its internal policy failures, such as by harping on India's failure to provide it all the water it needs through Farakka. Pakistan is also sensitive to water-sharing arrangements, but it has the Kashmir issue to divert public opinion for its policy failures. Sobhan (2000) argues that all countries in the sub-continent use such situations for internal political advantage. A regional approach would blunt their efforts to a large extent.

In India the voices of most people have remained unheard in the corridors of the water bureaucracy, which has dominated policy making. Even when by sheer coincidence the voices of stakeholders have received recognition, every instrument has been used to discredit them. The vociferous reaction of the GoI and the water bureaucracy to the critical WCD report of the Narmada project is an example. And this is happening in an age when there is virtual agreement that "development without people's involvement and development not meant for people cannot be a development" (Ravi and Raj 2006: 2).

Parliament is supreme in overseeing government's policies in a pluralist democratic society, but the Indian Parliament has been less effective in this area. One NGO found that one-fourth of the elected representatives in the Indian Parliament have criminal charges hanging over them and more than half of the members of the lower house are super-rich in a country where more than one-third of the population live below the poverty line. Indian political parties are also at fault, as there is a distinct lack of policy focus in their manifestos (Weiner and Kothari 1965: 7). Moreover, a large number of Indian political parties, besides being based on undesirable political leadership, are also based on linguistic, regional, religious, and caste affiliations (Weiner 1989: 179–222). In this environment it is not difficult for the highly organized dam industry to influence policy where the stakes are high.

Because of the exclusive focus on dam building in India, the biggest losers are the displaced refugees, most of whom lost everything they had including their dignity and pride (Singh 1997: 182–203). There is not only no reliable estimate of their numbers but also no genuine national policy for rehabilitating them. The UN had a long-standing policy on the rights of displaced persons caused by internal strife and government actions, but like many other UN policies it received little attention from water policy makers (Vincent 2001: 8–10; Martone 2006: 129–144).[1]

While this is broadly true with all displaced persons, the displaced persons from the relatively small projects remain the worst sufferers, as they do not receive outside support to highlight their plight, as was received by the Sardar Sarovar project displaced persons (Sharan 1997: 446–48). That may change, as the US Senate has moved to make water a top-priority issue in US foreign policy (Parker 2005), which has been alleged to be prompted by the abandoning of villages by water refugees in Iran, Afghanistan, and parts of Pakistan.[2]

If stakeholders had been directly involved in the policy development process, many ill effects of past water policies could have been avoided. It would have saved large amounts of taxpayers' funds in addition to keeping thousands from becoming destitute. The Arun III power project in Nepal is an example of how the grassroots participation of stakeholders in policy making not only saves money but generates employment and income for the poor, although World Bank backers of large dams opposed it. However, this would require that all parties, particularly the parliamentarians and the executive commit themselves to address the real issues, and refrain from creating commotion thus hiding the real issues. For example, Prajapati (1997: 693–94) highlights instances when the parliamentarians failed to do so while discussing the Supreme Court judgement on maintaining the height of the Sardar Sarovar dam at its present level.

The environmental effects of water-related projects and their impact on the poor and marginalized are predicted to be extremely high (Revkin 2007). These are increasingly being scrutinized by the powerful NGO community, both in the developed and developing countries. For example, the strength of the environment movement forced the Australian government to intervene and stop the state government from proceeding with the Tasmanian Franklin Dam project in Australia in the 1980s. Also, the Spanish government decided not to proceed with the large Ebro project to transfer water from the wet north to the dry south (Hooper 1995).[3] NGOs' important role in shaping public opinion by virtue of their work at the grassroots level has been reconfirmed at a recent international workshop on globalization, and that politicians would find it difficult to ignore such opinion in the days to come was reinforced (Pacific Institute 2005).

NGOs and community-based organizations therefore need to be central for a sustainable water policy, subject to two conditions: they must be self-funded to keep them immune to undue sectional influences, and they must be prepared to accept the compromises required to develop a balanced public policy. The Rio Conference in 1992 to deal with water questions failed partly because the compromises were insufficient, as many participating NGOs remained focused on a single issue.

The failure to involve stakeholders in policy development led to the current upsurge in inter-state water disputes in India, with each state acting for itself and demanding freshwater access that cannot be provided. Even the World Bank acknowledged this in a recent report when it acknowledged that "water conflicts are becoming endemic at all levels" (Briscoe and Malik 2006: xix). Coalition governments have only made the situation worse. If stakeholders were involved and apprised of the relative merits and demerits of policy options, they would generally be willing to support policies that can be sustained in the long run, even though some may hurt them, as the "Indian people have shown great ingenuity in working around a poorly governed water system" (Briscoe and Malik 2006: xvii). And if the possibility of misunderstandings, conflicts, environmental damage, and wastage is factored into policy from the beginning, with stakeholders participating in that process, such problems can be avoided or at least minimized. It is essential to understand the whole range of actors involved—their identities, interests, and loyalties, as well as the linkages and alliances between them.

This leads to a very important policy issue: the need to have a policy baseline. A baseline would help keep stakeholders mindful of long-term policy needs and direction, and prepare them to accept the good with the bad. It would alleviate uncertainty in the policy direction and avoid stink from harsh policy decisions that politicians are often reluctant to make for fear of their political costs. In India politicians are hesitant to address tough policy issues, and unpopular policies—irrespective of their merits—are usually frowned upon. A failure to charge an economic price for freshwater, a scarce natural resource, is an example.

A policy baseline provides directions to individuals and organizations for initiating actions at micro-levels. The baseline sets out guidelines to ensure that micro-level initiatives are not in conflict with the strategic policy directions. In a sense, this is something like a constitution that sets the parameters within which governments and the community are required to operate. This would also help promote economical agricultural practices. For example, it would be easier to argue that Punjab should not cultivate rice, or that Maharastra should not focus on sugar cane.

A policy baseline, generally speaking, makes it easier to address existing and emerging critical issues even through unpopular measures, as the stake holders get familiar with nature of such problems well before policies affect them. In a rapidly changing complex economic and political environment governments are often required to adopt unpopular measures at short notice. For example, should the industrialized nations eventually agree to remove or reduce subsidies from their agricultural products, it would be foolish to assume that they would not then look at the water management policies of developing nations, as many do not charge an economic price for irrigation water (Dyson, Cassen, and Visaria 2004: 367–68). As industrialized countries are increasingly introducing a non-subsidized pricing policy for irrigation water, it would be logical for them to challenge any subsidized service on the grounds that it creates an unequal playing field. Their capacity to influence global policy outcomes on economic and political issues should not be underestimated (Grynberg 1998: 11–56, 90).

A policy baseline also indicates long-term direction and offers policy certainty to stakeholders and consumers that would enable them to take initiatives to introduce new ideas and make necessary investments with some degree of certainty. In other words it would contribute to encouraging innovative measures in their respective fields of activities, e.g., investments for better drainage systems in irrigated areas and new water-saving practices.

Given India's poor financial situation, the need for a *total integrated approach* while formulating national water policy becomes a critical priority. Furthermore, given the regional, environmental, physical, ecological, and biospheric realities, water should not be used in a way that adversely impacts the communities in either the upper or lower riparian countries, both in the national and international contexts. That is the underlying principle of the 1997 draft UN Protocol. Most of the internal water-conflicts in India—e.g. between Tamil Nadu and Karnataka or between Punjab, Haryana, and Rajasthan or between Andhra Pradesh and Maharastra—are rooted in a failure to heed these basic principles.

Need for a Comprehensive Approach

Many earlier civilizations disappeared because the relationships between land, water, and the environment were not adequately understood, contributing to injudicious use of freshwater. But land and water were plentiful then, and people were able to move when either became unusable or scarce. Their failures cost them their habitat, but usually the consequences were not devastating since they could move somewhere else. Many tribals lived this way until the early to mid-twentieth century.

The demand for freshwater increased substantially in the post–Second World War period with rapid population growth, which, in turn, led to an increased demand for food production and other developmental activities. By the late twentieth century two things became clear: increased demand for water could not be sustained, and injudicious use of natural resources had wreaked untold destruction of the environment. If the demand for water and other natural resources is not checked and they are not used judiciously, the weaker nations in particular will suffer starvation, disease, and malnutrition in addition to uninhabitable environments.

As the supply of freshwater is outstripped by demand two types of conflicts are emerging: conflict between humans and other forms of life on earth, particularly animal life, as humans and animals compete for land and water; and conflict between humans over the same scarce supply of water, with the weaker missing out or marginalized. Water-related conflicts have increased sharply since the 1990s.

When conflicts arise over a rightful share of a basic but scarce resource, they affect every aspect of humans' existence, including their health, the economy, the environment, and their security. Water policy must take into account the interdependencies of human activities; no policy can be sustained otherwise. The importance of a total approach cannot be exaggerated. But truly integrated, holistic, and interdisciplinary planning that involves marrying land use and water use on the demand side and integrating all development activities on the supply side has not been seriously attempted in India (Iyer 2003: 71–72).

Water Policy in the Regional Context

A regional water policy is critically needed for four reasons. First, it is necessary to ensure that water flows from outside India remain steady and unpolluted. Failure on each count can be disastrous, as is the case in Mexico, a lower riparian country to the powerful United States, or the Netherlands, which suffered enormously with the accidental pollution of the Rhine caused by the Sandoz fire in Switzerland. Pollution in the waterways of China, Nepal, or Bhutan can be disastrous for India.

Second, there have been nine conflicts between India and its neighbours (Gleick 2003a), and there is little doubt that the water issue has kept some of these conflicts alive. The political importance of the entire region to the global power-players needs to be considered and understood, as the global power-players may find it useful to keep some conflicts alive by manipulating events to help their cause. A regional approach would limit their ability to do so.

China is integral to India's water policy because of its capacity to tap the Brahmaputra and Indus waters. In its quest for new sources of water, China will almost certainly consider harvesting water from rivers flowing through its territory. India needs Brahmaputra waters both for the development of the northeastern part of the country and to augment water flows in the lower Ganges. China holds some 48 percent of the Brahmaputra basin, and it would be difficult, if not impossible, for India to act unilaterally to divert Brahmaputra water. Their current friendly relations aside, India and China have not resolved their forty-two-year-old border disputes. It is difficult to predict the future direction of their relationship, particularly in the light of China's ever-expanding economic, military, and political power (Oster 2006; Margolis 2006; Shaplen and Laney 2004).

China's response to the water-management and other issues in the region will particularly be determined by a number of interrelated factors: first, how it reconciles its internal political pressures created by rapid economic growth with the domestic demand for more freedom and how its future economic and political relations develop with the United States, the EU, Japan, and resource supplying countries. The recent satellite test, notwithstanding Chinese leaders' insistence that their goal is to use such power for peaceful purposes, indicates that it is not thinking of the long-term good of the globe, argues Ms. Elizabeth Economy (2007), an Asia analyst. While this remains a strong possibility, it is also worth noting that high economic growth rates achieved by both China and India in recent years, and India becoming a member of the *nuclear club* may encourage both these countries to resolve water-sharing arrangements without any rancour (Prestowitz 2005).

Nepal is strategically important to both China and India. It is in political turmoil, and its internal political dynamics is likely to bring about major domestic political realignment in the coming years once the current political realignment firms. Depending upon the nature of this realignment, India's options could become limited if realigned forces play the China-India card to secure maximum advantage.[4] A regional water-management strategy would act as a buffer in the event that one or two countries destabilize the existing water-sharing arrangements.

The World Bank–brokered treaty between India and Pakistan on sharing the Indus water had been working well until India recently proposed construction of two new dams, which Pakistan objected to. One dispute has now been resolved through independent arbitration. Should China contemplate to harvest or make a claim on water from the Indus systems, this would require independent arbitration involving China, Pakistan, and India. A pre-emptive regional approach may provide a better chance to obtain a favourable outcome.

Third, a regional approach would also prevent smaller countries from misrepresenting India in international forums on water-sharing arrangements. Bangladesh and Pakistan, both of which have strong fundamentalist elements and lower riparian countries to India, often demonstrate a pathological animosity in their dealings with India. They also find it relatively easy to portray themselves as victims of Indian hegemony, or to portray India as taking an uncompromising big-brotherly attitude towards them. A regional approach would make it difficult for them to make such claims.

Fourth, there is no super-agency entrusted with the power to dictate and enforce international law to resolve water-related conflicts. Nor is there a regional organization with enough credibility to effectively address disputed issues in the region. Article 38 of the Statute of the International Court of Justice enumerates in order of precedence the sources of international law that the Court must use. Of the four major sources, only the Law of Treaties—the body of treaties and conventions ratified by governments—is binding; the degree of applicability of other sources of international law depends on the judgments of the judges and arbitrators (Cano 1989: 167–71). Any uncertainty that is inherent with the degree of applicability of other sources can be avoided with a regional approach to resolving water-sharing issues and that would be enormously beneficial to India.

It is impossible to know for sure how events will move in the future. But one thing is certain: water-related conflicts, whether with a powerful neighbour or with poor states riddled with internal conflicts, will not be unilaterally resolved in a world rife with cultural, ethnic, and religious sectionalism. The Islamic Jihad is only one source of it. The sectionalism needs to be handled with each country's long-term economic and political interests in mind (Norris and Inglehart 2004: 215–41). A regional approach would blunt some of the destabilizing elements.

India has been practicing bilateralism with its neighbours. Its insistence on bilateralism is understandable, but in international relations no position can be permanent. In practical terms, most discussions between countries are bilateral, but if keeping third parties informed reduces misunderstanding and nourishes constructive action, there is no reason to back away from wider consultation, even if it is informal initially (Varghese 1996). It should be central to a regional approach to water-management issues for India.

The task will not be easy (Frederiksen, Berkoff, and Barber 1993: xx–xxi), as lack of cooperation has characterized relations in the sub-continent—most countries in the sub-continent are conflict-torn, suffering from fragmented power-struggles of various natures. Good governance requires an all-around commitment, both within and in the region, not to pursue small chips of power by neglecting education, health, justice, and other basic needs of citizens (Newberg 2006) that must include guaranteed access to freshwater. It is, however, worth remembering that water practitioners in South Asia collectively acknowledged long back that there was not a single instance in the region where integrated water resource management was successfully implemented for the simple reason that sovereignty of the states do not necessarily coincide with the planning and management needs of such approach (Tortajada 2003: 130–31), although water sharing is a critical issue in all these regional countries. Hopefully, in a multitude of small ways, minor shifts in perceptions and new choices and dilemmas will facilitate this process (Rose 2005).

Multilateral and bilateral development agencies can play important roles in encouraging governments to adopt a regional approach to riparian water-sharing issues. As governments seek assistance for water-related projects, these agencies can directly or subtly encourage them to consider addressing the issues regionally. This can be done in two ways: by providing technical advice highlighting the merits of a regional approach, and by offering a financial carrot in the form of a concessional loan or grant should they agree to approach the issue from a regional perspective. However, past experience indicates that some agencies have been guilty of violating stipulated norms when putting pressure on governments. They need to ensure that their approach remains above board and creates no basis for criticism. Only then will they be able to play an honest broker's role in encouraging a regional approach.

Stakeholders' Participation and the Need to Establish a Policy Baseline

When the British colonial administration in India proceeded with the system of canal irrigation despite warnings from British engineers and soil scientists about its negative effects, the authorities allegedly found *scientific ways* to justify excluding local communities from administration of the system (Gilmartin 1994: 1137)! But without direct access to local knowledge, the colonial administration failed to notice the gradual deterioration in soil capacity, increased salinity, and other problems.

Post-independence water-policy makers did not shift from this *Raj* tradition of ignoring people at the grassroots who have immense practical knowledge gathered over generations (Sengupta 1985: 1919–38; Mitra 1996: A31–A37; Mohan 2003: 442–53). But grassroots contributions and participation to policy development are too valuable to be ignored (Priscoli 2004: 221–27). Priscoli argued that control of water is control of life and the way this control is executed is closely linked to the civic culture, a critical element in water resource management. For example, Australia suffers from devastating forest fires annually, causing enormous destruction of flora and fauna, property, and occasionally human life. The Aborigines used to control-burn forests to avoid such destruction, and in the process they increased land productivity. European settlers, without understanding the complex relationship between fire, land, and the preservation of flora and fauna, stopped this practice at their peril. Recent studies rediscovered the virtues of the Aborigines' policy, and governments at all levels now practise controlled burning of forests (Sneeuwjagt 2005). This confirms the importance of empirically based collective judgment in environment and water policies (Soltan 1996: 202).

But in India grassroots participation through elected representatives has not been effective, as the positions of the political parties have not played a major role in past election outcomes; nor do they now. Many political parties are primarily guided by naked ambition. This has helped create a powerful class of people who do not hesitate to exploit the resources of the state for their own benefit (Homer-Dixon 1996: 14–16; Seshan 1995: 13–71 and 264–70; Weiner 1987: 282). They take full advantage of feudal socioeconomic conditions, and the landed gentry have their own way of preserving their hold on scarce resources such as water (Mollinga, Doraiswamy, and Engbersen 2004).

Land reform is critical for stakeholders' participation in water policy development, particularly in countries where the rural economy dominates. Successful land reform was one of the key elements in the phenomenal economic growth in Japan, Taiwan, and South Korea. On the flip side, failed land reforms led to revolutionary violence in Russia in 1917, the overthrow of the emperor in Ethiopia, and armed struggles in the Philippines, Nepal, and many Latin American countries (Power 2005). In spite of this and notwithstanding increasing evidence that a more equitable distribution of farmland actually leads to better productivity, third world countries, including India, have been reluctant to implement rigorous land reforms, argued Dogra (1996: 2725). It is certainly one of the politically most difficult reform policies to implement, as China experienced despite its one-party rule (Hinton 1969).

To prepare a policy baseline with the direct participation of a majority of stakeholders is not easy, but it must be undertaken, because the distribution of water is part of the long-term management of the resource (Rijsbersman 2004: 87)[5] and policy-manipulation mostly occurs at that juncture. To minimize and overcome such possibilities the following mechanism is suggested to ensure fullest possible participation of all stakeholders in sustainable water-policy development in India, the most complex of all complex societies on earth:

- A high-powered task force to recommend a sustainable water-policy baseline for consideration—something that can only be developed by an independent interdisciplinary agency;
- It should commission research papers on all aspects of water policy and invite submissions from individuals, NGOs, community groups, and professional organizations on the long-term policy direction;
- It should invite anybody it considers important to present their views in person, and that should be done transparently; and
- Based on submissions and public hearings, the task force should prepare its draft recommendations. These should be filtered through further public participation, this time through public hearings in a limited number of regional centers, before final recommendations are submitted.

The success of this process depends on five elements:

- The task force members must be selected on merit alone so as to have the unquestioned confidence and trust of the community. They must be well-known for their personal and professional integrity;
- The terms of reference for the task force must not be prepared unilaterally by the government. It should be done through widespread consultation with stakeholders. Even then, given human nature, some people may not feel fully satisfied. To overcome any semblance of impropriety or dissatisfaction, the task force should be given authority to cover additional areas, should it feel it necessary;
- It must be given adequate resources to carry out its work and be allowed to perform its duties without government intervention; that should reinforce community confidence in the task force;
- It must be given the legal power to ask, at its discretion, individuals, organizations, and government officials to give evidence before it, including information on past policies and the reasons for their successes or failures. It is vital that past policies are clinically dissected to ensure that the same mistakes are not repeated; and

- The task force's ability to conduct its business transparently and to inspire stakeholders to participate in the process in a non-threatening and dignified way will be the key to the success of the exercise.

It is unrealistic to expect the task force to complete its work in less than three years. An important element of this effort must be to adopt a massive community education program to impress on everyone that water is a scarce natural resource and should be treated accordingly. A supply-driven approach and inefficient management of a key resource such as water are resulting in poor service delivery. Ultimately this results in immense suffering by the ordinary people, mostly the marginalized and those who are in the lower end of the society. If not for anything else, this situation necessitates reforms in the sector to ensure that people feel ownership of the policy and involved in the management of resources.

The task force's recommendations need to be well argued so that governments won't shelve recommendations they do not like and only implement those that suit their narrow political or sectional interests. To avoid any political conflict, the task force report should be considered by Parliament, not the government of the day.

It is heartening that despite the World Bank's many failures and its inconsistent policy approach on water, it now recognizes that community involvement in all aspects of irrigation, where more than 80 percent of water is consumed in developing countries, is the key to the success of water policies (World Bank 1995, as cited in McCully 1996: 25). The Bank in a recent publication frankly acknowledged that India is facing a "Turbulent Future" with respect to water and has suggested ten "Rules for Reformers." While scope exists to argue with respect to some of its suggestions, three suggested rules deserve particular mention, because they collectively summarize the complexity of the tasks, which must be undertaken. These are: "Water is Different"; "Reform is Dialectic, not Mechanical"; and "There are no Silver Bullets" (Briscoe and Malik 2006: 63, 67–68, and 70–71). On water-policy issues, the Bank does not have an unblemished record, but it deserves kudos for pinpointing these three critical issues!

Conclusion

Successive Indian governments have not assigned due importance to the need for a sustainable water policy. The policy documents are a hotchpotch of random thoughts, devoid of any meaningful direction or program. Water administration has been plagued by lack of foresight, corruption, inefficiency, bureaucratic bungling, and lack of proper coordination at all levels (Chapman 1992: 41). The casual attitude of the administration is evident from the government's failure to recognize water as a scarce economic natural resource and its failure to consider the issue in its entirety. Most importantly, water-policies take time to achieve the desired policy outcome, and no policy can succeed if the people, the principal stakeholders on freshwater issues, do not become a direct and active party to that policy formulation.

We have already noted that water is now a *Human Rights Issue* and deserves topmost priority from the policy makers. It is a human issue, and water problems we confront have a human face, but we have failed to make this a priority. It is time to take the necessary steps to prevent this needless suffering. In the new millennium it is also clear that global climate change threatens our water resources. Extreme weather, sea-level rise, and hotter and hotter temperatures threaten to alter water supply everywhere. The climate cycle is the water cycle, and attacking and adapting to climate change requires the focus of all of our political, community and business leaders.[6]

The government's immediate policy focus—the river-linking project—has infuriated environmentalists, riparian countries, some states, and a large number of experts both nationally and internationally. The Plata Basin (covering Argentina, Bolivia, Brazil, Paraguay, and Uruguay) experience shows that historical rivalries led some nations to interpret certain integrating actions as risky. Implementing international cooperation requires as an early step the establishment of an argument that should specify the objectives and the reasons for joint action (Trevin and Day 1990: 87–105). Such a mammoth project demands careful analysis, and all of its aspects must be debated and argued openly, with the full participation of all national stakeholders and cooperation of riparian countries. The task is not easy, but there is no alternative for all the mainland South Asian countries.

The feasibility reports for a few smaller canal-link projects along with state government comments on them have been made public. But feasibility reports have been consistently manipulated in the past under a cloak of secrecy, and the same trick has been pulled again. Until all stakeholders have deliberated on all aspects of the project and agreed to a well-considered and balanced policy baseline, and until the affected riparian countries are provided with detailed technical assessment reports of the entire project, it is sheer lunacy—environmentally, economically, financially, socially, and politically—to even think that the project is viable. Let's heed the concerns raised by a large number of highly regarded scientists and researchers, academics, environmentalists, NGOs, and social and political commentators, who have questioned the secretiveness and simplistic logic of project proponents and the haste with which the entire process has been conducted.[7]

Notes

1. The Asia Pacific Forum of National Human Rights Institutions 6th Annual Meeting at Colombo, Sri Lanka (2001), reconfirmed that "While the responsibility for the protection of internally displaced persons rests first and foremost with national governments and local authorities, the role of national institutions in ensuring that the State meets its obligations and that the human rights of internally displaced persons are protected and highlighted." http://www.asiapacificforum.net/training/idp/index.htm (December 18, 2006).

Also, the following statement issued by Ms Sagata Ogata (former) UN High Commissioner for Refugees on internally displaced reflects global thinking: "I believe the [UN] Commission [on Human Rights] in its work to strengthen the protection of the internally displaced must seek to bring about a convergence of refugee law, international human rights law and international humanitarian law. . . . [E]ach has a useful contribution to make to the protection of the internally displaced. . . . It is only through such convergence that the lacuna in the law can be addressed" (UN, Geneva February 9, 1994; cited in Cohen and Deng (1998: 71).

2. *Voice of America* English news (2005), March 17.

3. "Spain ditches Ebro river project"; http://news.bbc.co.uk/2/hi/europe/3817673.stm (June 19, 2004).

4. For an interesting discussion, see Emmot (2003: 56–76)

5. The Copenhagen Consensus Project was organized by the Environment Assessment Institute, Denmark, with the cooperation of the *Economist* of London. The project considered and prioritized a series of proposals for advancing global welfare. Papers presented at this conference were published in 2004: *Global Crises, Global Solutions* (Ed.). B. Lomborg; Cambridge University Press.

6. Summarized and abridged version of the *World Water Day Message* (March 22), issued by the (California based) Pacific Institute and its chairman, Dr. P. H. Gleick.

7. An open letter to the President, Prime Minister, the Task Force Chairman signed by fifty-eight experts raised many critical issues. These need to be openly considered so that existing doubts can be addressed (*EPW* 38/40: 4278–4279). This is just one of many representations made by a large number of experts suggesting an open and transparent evaluation of all aspects of the proposal. Also see Singh (2003: 4277–78).

APPENDIX A

Insurgency and Water: Three Cases

The Chittagong Hill Tribes Conflict in Bangladesh

The Chittagong Hill Tribes, known as Chakmas, consist of at least twelve tribes. Their unity is based on the adversity they have faced since the creation of Pakistan in 1947 and then the birth of Bangladesh in 1971, although the root cause of their dissatisfaction can be traced back to colonial days when the British deprived the Hill people of their means of production and sustenance. Hill tribes are "jhum" cultivators. They move away from their village to cultivate the land and return to the village after the harvest is completed. Land is the vital element around which the lives of the tribes revolve, as "jhum" cultivation and harvesting of forest products are their primary occupations.

The Kapti dam on the river Karnafuli was constructed as a multipurpose hydroelectric project in the 1960s to expedite economic development in the region. Therein lay the immediate root of the conflict. The dam submerged an area of about 400 square miles that included 54,000 acres of cultivable land, 40 percent of the district's total acreage. As a consequence, between 18,000 and 19,000 tribal families consisting of about 100,000 people were uprooted; they became both homeless and landless, and were left with no economic opportunities (Anti Slavery Society 1984, as cited in Bradnock 1992). The government's rehabilitation programs were scandalously deficient: they were inadequate and poorly designed; the corruption of government officials made the situation worse for the people adversely affected by the construction of the dam; and a blatant bias was shown by the Bengali officials in favor of the affected minority Bengali population (Zaman 1982: 75–80; Mohsin 1997: 102–7; Penz 2004: 83; Samad 2004). During the long conflict, about 100,000 tribals took refuge in India. This has had long-term political implications for India, as many believe that some sections of the disgruntled tribals have developed links with the Indian insurgents, the Tripura National Volunteers (Chari 2003: 24).

The Tribes' loss of their land kindled a discontent that was aggravated by the failure of the government to show the sensitivity that is required when dealing with two different cultures. This discontent was further aggravated by the marginalization of the Tribes people caused by the construction of the dam. A more subtle source of their discontent was a constitutional measure that abolished the "excluded area" status of the Hill district, with the consequent dismantling of tribal administration, the transfer of tribal officers to other districts, and the advice by the late Mujibur Rehaman to the tribal people to turn themselves into Bengalese (Barman 2004). This state policy of assimilation failed, as the Tribes people did not get fair treatment from the mainstream Bangladeshis. At the same time, new opportunities opened to settlers from outside the Hill district to settle there, and Chakmas were left with virtually no economic opportunities.

For the dam project to succeed it was essential that groundwork be meticulously prepared to assure its acceptance by the original inhabitants. A few such preconditions were a carefully worked out plan with the direct and active participation of the affected population; the establishment of an administrative arrangement that had the confidence of the affected population; and maintenance of the political status quo. But these were evidently not the objectives of those involved in the project, and consequently the assimilation of the Hill Tribes within Bengali nationalism was promoted. Hence the emergence of the insurgency movement, which got India entangled when the Tribes people fled to India from the conflict zone.

An agreement between the tribal leadership and the government of Bangladesh was signed, but it failed to placate a significant section of the community. A survey of the tribal population revealed that 69 percent felt the project had adversely affected them economically and reduced their ability to feed themselves (Dogra 1986).

Unless the mind-sets of officials change and the ordinary tribal people are convinced that the government's policy is genuine, the insurgency movement will continue. Some things are within the government's capacity to deliver, others are not. Although the agreement is designed to have a special 'hill people's representative' to deal with the government on issues relevant to them, the current occupant of this position is himself a migrant settler. This has great symbolic significance and is unlikely to raise confidence in the minds of the tribal population. Furthermore, while the government can provide training and financial assistance, this is unlikely to deliver jobs to the displaced people in the same sector, as even nationally such opportunities are few.

Another danger arises from the geo-strategic importance of the region and the belief that the region is endowed with valuable mineral resources. This makes the region vulnerable to political intrigues that may not favor the tribal population. In an intra-nation conflict, strategic location and the lure of resources more often than not attract external actors who can aggravate the situation (Chakma 2004).

The Tamil Tiger Conflict in Sri Lanka

The Mahaweli Basin development program is the largest construction project ever undertaken in Sri Lanka. The program goes back to the early 1950s, when the World Bank, UNDP, and FAO were involved in various aspects of plan preparation, evaluation, and project design. The project was designed to irrigate 900,000 acres of land, directly affecting over half a million people, and to generate 500 megawatts of power.

At its peak the plan was absorbing 6 percent of the country's GDP, 17 percent of the total public expenditure, and 44 percent of the public investment expenditure (Athukorala and Jayasuriya 1994: 115–16). The project contributed to ethnic conflicts already fueled by actions of the Sinhalese government that disadvantaged the Tamil population. Dam proponents ignored both environmental and social risks and proceeded with construction (Alexis 1985), fueling a fire that has now engulfed the country, with dire effects on the entire island community, economically, socially, and politically.

The project provided the government with an opportunity to settle Sinhalese on lands that the Tamil community regarded as traditionally its own, thus changing the ethnic balance and worsening the relationship between the island's two main communities (Bradnock 1992: 50; Swan 1987: 298–390; 422–35). The Sri Lankan establishment saw the project as a golden opportunity for diluting the domination of the Tamil population, who became increasingly frustrated (Gunaratna 1988: 32–33; Scudder 1990: 1–5). This initiative and the discontent that had existed for a long time were the immediate reasons for the subsequent armed struggle launched by the Tamil Tigers, who eventually demanded an independent "Tamil Land."

The Tamil population consisted of two distinct socioeconomic groups: the elite class that occupied senior government positions, which was by nature conservative and which was unsuccessful in achieving the objectives Tamils hoped to achieve; and plantation workers and others who were less well-off. As the minority Tamil population felt increasingly alienated from the mainstream, the young generation in particular started to lose confidence in the conservative Tamil leadership based in Colombo. The struggle has directly caused the deaths of a large number of people to date. Pearce (1992: 155) estimated that about 65,000 died by the beginning of the 1990s alone.

The Chico River Basin Development Project in the Philippines

The Chico River Basin development project in the Philippines has created considerable distress for Filipinos in general and even more for the people who live in the basin. It is a classic example of how an insensitive approach by government and business can provoke inhabitants of targeted development areas. In this case the inhabitants became increasingly militant (May 1997).

The Bontoc and Kalinga people in the basin opposed the government's proposal to construct a number of river dams, which they feared would eventually flood their ancestral land (Goldsmith and Hildyard 1984 vol. i: 29). The affected communities joined together in a compact. The government confronted the group by declaring the area a "special development region," and a government-sponsored organization known as the Private Association for National Minorities was set up to represent the interests of non-Muslims in the dispute. The association was actually established by a wealthy businessman with the hidden agenda of plundering the huge forest resources in the region, either for himself or for cronies of then-President Ferdinand Marcos, who had given him the opportunity to get involved.

Gradually the resistance movement attracted support from progressive political forces that were becoming increasingly vocal in their opposition to Marcos. The inhabitants themselves articulated their position this way:

> Water eats up land. Slides will occur, caused by dammed up waters. There will necessarily be heavy silting and the water level will rise. . . . The President says the dams are for the development of the whole country. But what about us? Others will be developed, but we—we are expected to go up to the mountains and eat grass (Adams 1988).

As the movement started gathering support from progressive political elements, the Communist New People's Army became increasingly active in the region. That provoked the government to deploy its army, including the Civil Home Defence Force, which was more famous for its poor discipline than for its professional efficiency (May 1997). The Bontoc and Kalinga people, along with the Isneg and the Tingguian, fought desperately to stop operations by a logging company that had been granted concessions for some 20,000 hectares. They fought government officials and prevented them from carrying out their surveying work for the preparation of the dam site. The government then intensified its military operations, provoking an international outcry. In the meantime Marcos's corrupt activities were galvanizing internal opposition to his rule and stirring up international opposition as well. Before being deposed, he was forced to postpone the scheme, as he had no alternative after the World Bank withdrew from the project in 1975 (Khagram 2000: 87 and 96).

The movement won the anti–Chico dam struggle, which encouraged activists to widen the scope of their struggle and boosted their confidence. Their fight against a Marcos-crony company called the Cellophil Resources Corporation was also successful. The organization now rightly claims that: The two struggles dramatically demonstrated the people's decisive stance to fight for their rights and their ability to muster widespread national and international support. In the face of the open fascist rule during martial law, this even meant resorting to armed resistance especially as tribal communities are traditional warrior societies (Bolinget 2004).

Postscript

These three cases show how governments manipulate their policies to achieve hidden objectives. In each case the local inhabitants were virtually forced to resort to methods that they had never previously adopted. In the process, long-standing trust between the rulers and the ruled was destroyed. It has yet to be restored.

In the Sri Lankan case, the driving force behind the government's determination to exploit minority groups and minority regions was money. The profits allegedly found their way to rather dubious accounts. In the Philippines, tribal areas were exploited for decades by non-native loggers, miners, ranchers, and farmers in cahoots with government authorities in Manila, while the locals increasingly lost their livelihood (Brown 1997).

In India, although the country has been faced with a number of serious insurgency movements since independence, their cause, generally speaking, has been the government's failure to respond to the genuine grievances of the affected or concerned populations. Major problems started in the 1960s. Since then, at the state level, ethnic politics assumed sharper overtones in places such as Tamil Nadu, Nagaland, Mizoram, and Punjab. In these states the demands of the dominant ethnic community have led to direct confrontations with the central government at times. Since the 1970s, two seemingly contradictory processes appear to have been occurring: the sharp trend towards centralization and homogenization by the state; and a gradual decline in the authority of those in positions of power and a loosening of the grip of the national leadership. There has also been increasing resort to populist politics (Phadnis and Ganguly 2001: 367).

As a consequence, crucial decisions about the allocation of resources have come to be heavily influenced by political considerations rather than by sound technical and development criteria. The result is widespread graft and corruption (Sharma 2003). There has been a diminution in the legitimacy and effectiveness of the state caused by a centralized administration that is mostly based on either personal or party loyalty or obedience to a monocratic executive (Rudolph and Rudolph 1987: 84).

In this environment one can easily imagine how the involuntarily displaced—estimated to be somewhere between 20 and 50 million—or the warring states looking for a larger share of water from the interstate rivers will respond as the situation evolves.

APPENDIX B

A Template for an EIS

This template is primarily based on reports published by the South Australian Department for the Environment (1980), South Australian Department of Housing and Urban Development, Environment Australia (1997), World Bank Technical paper No. 139, and the UN World Water Development Report (UNWWDR: 2003). This should be taken as a guide only, and must be modified to suit the specific requirements of individual country- or sector-specific projects.

The UNWWDR strongly recommends that freshwater be managed in an integrated fashion to optimize its benefits. Understanding the environmental effects of water withdrawals and releases, the construction of storage dams, distribution channels, and the like is essential to the development of an integrated and sustainable freshwater policy. These considerations set the parameters of all EISs, whose scopes are also significantly conditioned by governments' resource management and related policies.

The river-linking project is a huge undertaking, possibly unparalleled in the history of freshwater management anywhere in the world. A critical issue is that the characteristics of no two river basins are the same, and hence interfering with their natural flows will have different effects on their basins. Although the World Bank has established procedures to specify these characteristics, many believe that the Bank's procedures have not been consistently followed (Munir-uz-Zaman 1987). It is essential that all assessments be undertaken by people with the unqualified confidence of most stakeholders in India and other affected riparian countries.

The UN has identified five critical aspects of managing freshwater resources besides environmental issues:

Economics
Allocation
Accountability
Covering costs
Financial resources

An EIS of the river-linking project should take into account the following matters:

Freshwater availability: surface water, groundwater regime, hydraulic balance, effects on drainage and water channels, flooding, and sedimentation;
Effects on land and land use: land capability, soil resources, and effects on soil structure, mineral resource exploitation, tectonic activity, and other geological features;
Effects on flora and fauna: environmentally sensitive areas such as wetlands and marshes, environmentally and culturally sensitive forests (particularly in indigenous and tribal areas), biodiversity, and animal habitats' nutrient-cycling;
Effects on human activities and institutions: social structure and institutions, cultural heritage, physiological and psychological well-being, and economic activities.

Within this broad framework, an EIS should contain the following elements, which should be made understandable to all stakeholders:

Objective of the proposal;
Background to the proposal;
Description of the proposal;
Description of alternatives, if any;
Environmental impact:
    based on existing knowledge
        based on changed circumstances likely in the foreseeable future (e.g., effects of climate change)
    during the construction phase of the project: short and long-term
    during the operational phase of the project;
Proposed environmental safeguards;

Monitoring and review;
Public participation;
Sources of information used in the assessment;
Decision-making authority and the legal basis on which decisions are made.

These general guidelines, which need to be adapted to the specific project, should cover the following:

Abiotic issues, including terrestrial, aquatic, and atmospheric;
Biotic issues, including terrestrial, inter-tidal, and sub-tidal;
Human issues, including population, land use, facilities, infrastructure, and activities.

The social and economic aspects that should be included in the assessment are:

Demographic characteristics that may be positively or negatively affected by the project;
Factors such as the number of displaced people whose economic status might be raised by relieving them of their dependence on land and transferring them to growing economic sectors;
Regional effects on employment, land use, employment, trade, commerce, and community.

There are other matters specific to the region and individual basins that may require detailed assessments that have not been included in the list. These need to be considered as well.

APPENDIX C

Sociopolitical Environment in Islamic Republic of Bangladesh, People's Republic of China, Royal Kingdom of Nepal, and Islamic Republic of Pakistan

An understanding of the sociopolitical environment of the riparian countries is critically important, as almost invariably countries' external policy approaches are determined by their internal policy needs. A regional approach to address a freshwater sharing arrangement in the sub-continent would not be possible, although highly desirable, until the internal sociopolitical situation in each of these countries aims towards this. In this regard, the Royal Kingdom of Bhutan is skipped because the Monarch is politically suave and has already taken steps to ensure that governance practices meet national and international expectations in terms of policy approaches to riparian water sharing and managing issues.

People's Republic of Bangladesh

Bangladesh came into existence as a secular state riding on the back of "Bengali nationalism." Mujibur Rehaman, founder of the Awami League and the first president of the country, was killed in a military coup within three and a half years of Bangladesh's birth in 1971. Since then the country has spent fifteen years under military rule. Democracy was restored in 1990, but the political undercurrent has remained volatile all along.

Many political commentators believe that the "emergency" declared in February 2007, in reality, is also backed up by the Army. They believe that in the current international political environment direct military takeover runs a risk of being castigated by the global civil society on Human Rights grounds, particularly by the U.S. Congress and the Commonwealth "and that's why they're very keen to have the façade or the patina of a civilian administration behind which they are thoroughly calling the shots" (Fair and Ganguly 2007).

Subsequent declaration of emergency and other events put credence to this view. Charge-sheeting the former prime minister Hashina on murder charges along with her other political colleagues and debarring her from returning to Bangladesh from her overseas trip or sending another former prime minister Khaleda into exile with the promise that her detained eldest son on corruption charges will be allowed to join her soon indicate an uncertain political environment. A South Asia and Terrorism expert at the US Institute of Peace argues that "the problem is once this process of crackdown begins, it is very difficult for an organization like the army to stop." One should also remember that *there are those* (both regionally and internationally) "in the intelligence community who believe that Bangladesh is the next battlefield for political and even militant Islam" (ibid.). The political uncertainty thus created will not contribute to addressing the water-sharing issue constructively (*Economist* 2004: 67–68).

Since the murder of the founder President Rahaman, General Zia assumed power and established the Bangladesh Nationalist Party (BNP). His government led to the adoption of a non-secular Islamic identity. His loyalty to the military contributed to the military's emergence as a powerful political force in the 1980s.

A major difference between the Awami League and the BNP is that the freedom fighters who mostly form the basis of the former subscribe to radical and progressive views on national and international issues (albeit in a limited way), while the military, the base of the BNP, has retained a largely conservative and anti-Indian (Islamic) orientation. Also, the country's political history and tradition was not congenial to the growth of a true pluralist-democratic system because of prevailing socioeconomic conditions (Ahmed 1997: 116).

Bangladesh is economically poor, with one of the highest population densities in the world (see table 4.7, page 80). It shares fifty-four rivers with India and has plenty of water, but suffers from regular floods and from water shortages in the dry season. It is primarily an agricultural country and cannot sustain itself without ready access to freshwater. Its groundwater is heavily contaminated with arsenic—something that even the World Bank failed to realize until it became a serious health issue. Arsenic has already penetrated into the food chain.

The size of the Bangladesh economy is smaller than any of the world's top 500 companies.[1] About 78 percent of the population lives below the poverty line. The richest 20 percent receives about 43 percent of the total income, and the poorest 20 percent receives about 9 percent.[2] The government believes that food insecurity is manageable, but the booming population is a major obstacle to overcoming poverty.[3] Bangladesh is one of the world's six countries that accounts for half of the world's annual increase of 77 million people (Chamie 2004).

Its economy is largely based on agriculture, although during the last three decades it developed a garment industry that became the largest foreign exchange earner for the country. Seventy percent of the industry's three million workers are semi-skilled rural female migrants. The industry prospered on the back of a special export quota that is no longer available, as the industry no longer qualifies for special dispensations because it does not use one hundred percent indigenously produced material. This is having devastating effects on the economy (Grynberg 1998: 6, 11, 38, and 41; Gounder and Xayavong 2001: 5; Read 2001: 30–31; Arthur 2001: 603). One can argue that the country has paid the price of globalization (Rahaman 2004) and that Bangladesh has not prepared itself for it.

Bangladesh has had a severe law-and-order problem since its birth. Its governance practices have been questioned by many, including one former president and the British envoy to Bangladesh. The tribal-led insurgency in the Chittagong Hill district is also a part of its law-and-order problem. The Awami League government reached an agreement with the tribals that would allow them to maintain their centuries-old separate cultural and religious identity. The BNP did not support it, however, and sporadic clashes continue. The region is now a heaven for criminal activities, including murder, arms smuggling, and kidnapping.[4]

Transparency International found Bangladesh to be the world's most corrupt country. In trading with India, it officially imports goods worth over $1.1 billion a year and exports only $22 million. Informed commentators observe that illegal border trade is the reason for this huge discrepancy (Transparency International 2003: 160).

From time to time the government has used the Farakka Treaty to bolster its domestic political position. The treaty stipulates that should the flow in Farakka fall below 50,000 cusecs in any ten-day period during the lean season, India and Bangladesh will enter into immediate consultations to make adjustments in the treaty. Bangladesh accused India of violating the treaty when the flow remained below the stipulated level for a few days, and wanted India to assure a flow of 35,000 cusecs irrespective of the actual flow (Shiva 2003: 66).[5] Also, even after confirming at one international forum the need for improved regional political relations in resolving water-sharing arrangements,

Bangladesh raised similar issues at another international forum immediately afterward, fully aware of India's opposition to raising bilateral issues in a multilateral forum.[6]

Population pressures, loss of habitats caused by the rising sea-level, and lack of economic opportunities for Bangladesh's large unskilled labour force are forcing many to cross the porous border illegally to India (Chari 2003: 33; Hussain 2003; Hussein 2004). Compounding matters, since 1947 many Hindus have fled Bangladesh to India (Chapman 1995: 20). In 1947, 47 percent of the East Pakistan population was Hindu; the figure is now less than 10 percent. The way the 1947 Radcliffe Commission divided districts, villages, farmlands, water, and property has also contributed to this problem (Strauss 2002: 143).

While illegal immigration is creating significant social, political, economic, and security problems for India, Bangladesh has chosen to virtually ignore this issue, leading to an increase in tension between the two countries.[7] India's unemployment problem is severe. The migrants compete mostly in the informal sector, which is resented by the locals, similar to the situation in twenty-two European countries, Australia, and New Zealand (Gijsberts, Hagendoorn, and Scheepers 2003: 242). Research shows that the number of migrants entering India from Bangladesh will continue to rise and have increasing political, social, and economic implications for India in the next few decades (Dyson, Cassen, and Visaria 2004: 6).

After the partition of the sub-continent in 1947, the Pakistani leaders, including Bangladesh's founder, Mujibur Rahaman, started campaigning for extra land from India to accommodate Pakistan's burgeoning population in the east (Bhutto 1969: 163). Such campaigns have subtly fed mistrust at the community level in India. India cannot address this issue on its own, as a UN Commission for Trade and Development (UNCTAD: 1989: 39) found for African countries while examining similar issues in Africa.

India accuses Bangladesh of harboring Indian insurgents and offering them training facilities. Bangladesh's alleged failure to take effective action in response has also contributed to a political environment that fuels mutual distrust, and is a factor in the failure to address water-sharing issues in a wider regional and developmental context.

Some Bangladeshi academics and senior government advisors may have also unwittingly helped create this environment. A British academic commented that very few writers in Bangladesh were able to refer to the Farakka Barrage without repeating the same list of alleged effects on the country, including "desertification," a phrase normally used to describe an increase in desert margins, which is clearly not happening in Bangladesh (Hussain 2003: 139; Chapman and Thompson 1995: 187). Another academic argued that the proposed river-linking project would drastically change the ecosystem of Bangladesh, wreaking havoc that would virtually destroy its lucrative harvest of the Clupeidae Tenualosa Ilisha fish (considered a delicacy by Bengalese), with serious economic consequences.[8] He is not wrong, but he did not mention that the same is equally true for India, as the Ganges delta through which this species enters sweet water from the Bay of Bengal is common to both countries.

Myth propagation does not, of course, help either Bangladesh or India. Their conflict over Farakka and the barrage's effects on dry season flows cannot be denied. But for Bangladesh to scapegoat India for all other environmental ills and manifold additional problems is wrong and counterproductive (Chapman 1995a: 184). It has to put its own house in order, and the international community has to take active interest in ensuring that it cultivates and practices democratic values assiduously (Fair and Ganguly 2007). This undoubtedly will contribute in resolving the water-related disputes between these two countries.

People's Republic of China

China also suffers from water shortages. Some 2,241 CM of freshwater are available per capita there, compared to 1,878 CM in India. China's annual freshwater withdrawal of 525.5 BCM is higher than India's 500 BCM. It withdraws 18.6 percent of its freshwater resources; India withdraws 26.2 percent. Seventy-seven percent of China's withdrawal is for its agricultural sector, compared to India's 92 percent. The yearly rise in China's water consumption rate is low compared to other countries at a similar stage of economic development (Gleick 2003: 191).

In some areas of China the situation has reached a critical stage. For example, since 1985 the Yellow River has failed to reach the Yellow Sea for part of almost every year. Sometimes it does not even reach the last province it flows through en route to the sea. A few small rivers no longer exist, and almost 1,000 lakes have disappeared in Hebei province alone. The water table is falling by two to three meters a year in the north China plains.

China's approach to freshwater management has a better policy foundation than India's. It has adopted a basin-wide approach to data collection and to the planning and management of water of its main rivers, although a few structural deficiencies exist: it has not yet adopted the principle of the separation of regulatory and operational functions, for example (Frederiksen et al. 1993: xviii). The negative impact of this is clear: In only six of its twenty-seven largest cities does drinking-water quality meet government standards and the groundwater in twenty-three of those cities does not meet the standards either (UNEcoSoC

2004). India has hardly any basin-wide approaches with respect to its major rivers.

To meet its increasing demands for water and energy, China proceeded with the Three Gorges Dam on the Yangtze in spite of strong global opposition (White 2004). The dam displaced some 11 million people, directly affected the lives of 20 million, and destroyed 62,000 acres of farmland, 13 major cities, 140 large farms, and a countless number of villages (Fisher 1998; Jhaveri 1988: 56–63). An independent evaluation challenged many findings of the official evaluation report that supported the dam. Not surprisingly, the findings of the independent evaluation were rejected by people associated with the initial assessment (Barber and Ryder 1993: 22–32). Unfortunately the European governments actively participated in the project and reaped commercial benefits, although the United States opposed it on human rights grounds (Shoulders 1998, as cited in White 2004).[9]

However, China took a different position with respect to the Nu River project in western China. This project was designed to harness hydropower through a thirteen-stage dam that would have destroyed World Heritage–listed areas of unique biodiversity and displaced an estimated 50,000 people, mostly ethnic minorities along the river.[10] China's flexibility this time was a surprise, because China ignored all opposition earlier on such issues and because it was desperately short of energy. China may have taken this position because it wanted to show the world a prestige-enhancing "Green Olympics" in 2008 and did not want to confront the (Islamic) tribals who were strongly opposed to the project.

China has its unique way of dealing with contentious issues to suit its national objectives. It showed its unique style when it annexed Tibet in 1950 (Laird 2002: 128 and 271). It has adopted a flexible region-specific policy within the country, but takes extra care to make sure that the patronage element of the policy does not become prominent (Naughton 2004: 253–95). It is clear from its flexible approach to its much-maligned population policy, as it permitted significant ethnic, spatial, and temporal variations in places such as ethnic Tibet and Xinjiang (Huang and Yang 2004: 193–225).

A clear example of China's policy of putting its national interest first is its decision to remain an observer in the Mekong River Commission even though 44 percent of the river flows through China. This strategy enabled China to use the Mekong waters while ignoring international concerns and monitoring the commission's work. It has already constructed two dams on the river; two more are under construction, with another four planned. These have created economic and environmental problems for the lower riparian countries (Pearce 2004).

China's unilateral acts on the Mekong River violate two of the eight international substantive rights drafted by the UN Sub-Commission on Human Rights and the Environment. China has not signed the draft UN Protocol on Non-Navigational Use of Shared International Watercourses, presumably because it would restrict China's ability to ignore international concerns and protocols.

A regional approach to water-sharing arrangements in the sub-continent cannot proceed without China's participation. However, past events indicate that China would participate only when its national interest is served, and even then its participation would be on its own terms. Issues of such international or regional importance can only be dealt with politically within the UN framework. The recent international events such as the invasion of Iraq by Coalition of the Willing make one wonder whether it would be possible to come to such arrangements without the expressed agreement of powerful nations and whether those nations and their allies will agree to the arrangements, which would require them to part with some of the privileges they have been enjoying.

As China becomes stronger economically and politically, its approach to regional issues such as sharing riparian waters with India will be determined by its political and economic interests and how much it would be able to accommodate the interests of other regional countries.[11] From its past actions it is clear that China takes a long-term view of its national interest in addressing regional and international issues. To take one example: it is reportedly building defense-related infrastructure in Thailand and Pakistan and contemplating deepening its relationships with the Maldives, Bangladesh, and Myanmar to gain access to their military infrastructure to protect its sea lanes in the event of a conflict in the South China Sea or Taiwan Straits.

China is also actively involved in the Shanghai Cooperation Organisation, which it helped form. This is another means to protect its political and economic interests in the wider South-Central-Asian-Caspian Sea region in the post–Cold War and post-Taliban periods (Bakshi 2004: 285 and 303).[12] Potentially this region has huge reserves of oil, hence its political importance.

Despite the many meetings and discussions China and India have held on their border dispute, it has yet to be resolved, even though China has settled its border disputes with other countries including Nepal. Even with current friendly relations they have not resolved their forty-two-year-old border disputes and some believe that they still distrust each other (Hoge Jr. 2004: 3–4). In the case of Nepal, China did forgo claims on ten villages after initially claiming eleven. The issue needs to be seen in the context of China and India's political rivalry and potential economic competition between them. Trade interests often act as a buffer in difficult situations between countries. But trade between these two countries in the recent past is negligible, although scope for a substantial increase in trade exists. In 2003 its value reached $5 billion, which was only slightly more than 10 percent of the value of the trade between China and eleven Southeast Asian countries (Chung 2004: 117–18, 170).

China's focus on achieving rapid economic growth and protecting its security in a unipolar world may encourage it to adopt a more accommodating approach to regional issues (Ferdinand 2002: 130–45; Yahuda 2002: 102–13). Should this approach prevail, a regional orientation on riparian water-sharing arrangements is likely. On the other hand, China may consider it politically advantageous to keep some contentious issues with India alive. In that event a regional approach to addressing riparian-water management issues is less likely. There are some, however, who believe that China's approach to India is an enigma (Chellaney 2004; 2005).

China's unconventional approach is the natural outcome of restricting domestic political debate. In its drive to achieve rapid economic growth, China did not hesitate to marginalize farmers—the backbone of the Chinese revolution—in spite of their protests, by dealing with them in a typically Chinese way.[13] China's demonstrated capacity to ignore or at least bypass the spirit of the international conventions and protocols leads many to believe that current international water law may be unable to handle the strains of ongoing and future problems for riparian countries (Boyle 1996: 48; Dellapenna 1999; Gleick 2000: 213). A regional approach to water management in the sub-continent may not happen until the key players fully commit to working towards that goal.

Kingdom of Nepal

Nepal, a buffer state between India and China, is an upper riparian country for a few Himalayan Rivers and streams that flow to or through India. Only 17 percent of Nepal's land is suitable for cultivation, though a little over 90 percent of the population is engaged in agriculture—an economic environment that guarantees the abject poverty of the masses.

Nepal shares with India about 9 percent of the Ganges-Brahmaputra-Meghna Basin, and a number of the Ganges tributaries flow from Nepal. All Nepalese rivers feed the Ganges, the lifeline of India's northern and eastern heartland that has a total runoff of about 200 CKM (Upadhyaya and Sapkota 1985). Nepal currently withdraws only about 14 percent of its freshwater, of which 99 percent is used for agriculture (see table 1.2, page 11). The country does not suffer from freshwater shortages nor is it likely to in the future.

Nepal has hydropower potential of around 80 Giga Watts (UNWWDR: 2003: 262). It successfully opposed the World Bank–initiated Arun III multipurpose dam project and opted for small dams. The production-cost difference per KW between the small and mega dams is enormous: $700 to $5,000 per KW, with a substantial multiplier benefit for the national economy (Bell 1994: 113–15; Pandey 1994). The Indian focus on building large dams jointly with Nepal may become a contentious issue, should Nepal's political environment change.

The effects of climate change on Nepal's environment could be enormous, and that needs to be factored into the water-policy debate. The UN warned Nepal in 1998 that twenty big glacial lake expansions could trigger a huge loss of life and property. The Tsho Rolpa glacial lake near Kathmandu, which had an area of 0.23 SKM in 1950, now has an area of 1.7 SKM. There are some twenty-six potentially dangerous glacial lakes in Nepal.

Nepal has a long history of autocratic rule and poor governance. The situation did not improve much even after the first multiparty elections in 1991. The major political parties became famous for infighting and corruption, resulting in twelve governments in thirteen years, which fueled social discontent (Dahal 2001: 93). The police and judiciary remain virtually beyond the reach of ordinary people (Sherpa 1994: 9 and 72), and the poor have little chance to obtain justice.

All that has contributed to the Maoist insurgency that has provided the people with some hope of change. The rebels are not actively opposed by the population, despite their strong-arm tactics raising money for the movement (Bray, Lunde, and Murshed 2003). The king's proclamation of a state of emergency in February 2005 only made the situation worse.

The GoI has been concerned that this conflict could provide a psychological boost to some sections of the Indian insurgency movement, which has a similar political orientation. The Maoist has now joined the interim government, but has not denounced the arms-struggle to achieve their long-term political objectives. They have temporarily suspended their arms-struggle and deposited their arms under UN supervision for safekeeping.

The situation in Nepal is critical for India's freshwater policy, economy, and security. China is also taking a keen interest in Nepal's internal political development; its interest could influence Nepal's approach to India on water-sharing arrangements. China has already heavily invested in Nepal's infrastructure, which includes roads of strategic importance (Nath 2004: 65–66).

Disagreements on water sharing in any new political alignment will certainly be shaped by such investments.

Nepal's is constrained by its size and location. It needs to protect its interests by adopting a cooperative policy with its giant neighbours in managing its water resources and by ensuring that all dealings are transparent and benefit its citizens. Should there be a change in the internal political chemistry of Nepal, some or all of the existing water-management treaties with India may need to be renegotiated (Khadka 1994: 344–94). A regional approach would contribute enormously to achieving these objectives.

Islamic Republic of Pakistan

Pakistan was carved out of India in 1947. A large part of it is arid and deficient in rainfall, which varies between over 1,200 MM and less than 100 MM. Evaporation rates are extremely high, varying from 1,300 MM in northern Punjab to 2,800 MM in Sind. This causes huge delivery losses in irrigation systems, estimated at 40 percent on average (Ghassemi, Jakeman, and Nix 1995: 371–78). Although the aquifer is believed to extend 300 meters in depth and over most of the area, overexploitation without adequate recharge is creating problems, as the water level is receding rapidly, resulting in the drying up of an increasing number of wells. The country is not crisscrossed by rivers or waterways, and options for augmenting the freshwater supply are limited.

Pakistan owns about 56 percent of the Indus basin and shares the Indus system with India. The 1960 agreement on sharing the Indus River systems was hailed by many as a model. Pakistan's water crises are deepening, as Indus water flows often fall so low that when the stream enters the Arabian Sea it has been reduced to a trickle. Water tables have also been steadily falling in Punjab, Baluchisthan, and Frontier Province in the northwest. People from water-scarce provinces have already started moving out of many villages. There are indications that by 2010, Quetta, the capital of Baluchistan, will have exhausted its supply of water.[14] Pakistan's birth rate is more than forty per thousand and shows little sign of falling (Bradnock 1992). The demands created by this large increase in population are putting additional pressures on limited water resources.

A stable political environment has eluded Pakistan. Its founder, Jinnah, died within thirteen months of the country's birth and the first prime minister was assassinated within four years of its birth. Pakistani politics have remained closely linked to the Pakistani military, either de jure or de facto. Almost all Pakistani prime ministers—thirteen of them in fifty-seven years—have been humiliated and turfed out by the army or army-backed presidents.[15]

During the first twenty-five years of its existence, Pakistan adopted three permanent and four interim constitutions. A new constitution after each coup became the norm. The military is now one of the contenders for power (Rizvi 1998).

Pakistan's strong support to the Taliban regime in Afghanistan contributed to the growth of a strong Islamic fundamentalist constituency inside the country. Many consider Nawaz Sharif's overthrow as prime minister the result of the growing displeasure of the fundamentalists and the military when Sharif announced under international pressure a unilateral moratorium on nuclear testing and pledged at the UN to sign the comprehensive test ban treaty. He was also accused of complicity in the US missile attack on bin Laden's camp in Afghanistan.

Musharraf's alliance with the United States and campaign to wipe out fundamentalists within Pakistan represent an about-turn in policy. Those people who were encouraged by the Pakistan government in pre-Taliban days to fight the Soviets in Afghanistan felt the government had now turned against them (Norton 2004:131–32). Pakistan's crackdown in tribal areas thus has created a new militancy whereby young tribals do not want to talk peace and are vowing to fight to raise the banner of Islam. For the first time in Pakistan's history, religious parties made significant gains in the 2002 general election by exploiting strong anti-American feeling.[16]

Musharraf cemented the military's dominance in Pakistan and his own position by establishing a military-dominated National Security Council in 2004 and then by changing his own constitution to allow him to hold two offices simultaneously: chief of the army and president. He pledged to give up his dual role by the end of 2004 in exchange for allowing Pakistan to re-enter the Commonwealth. His volte-face raised strong opposition inside the country, although a few of his international backers found excuses not to oppose it directly. Musharraf's ethnicity makes him insecure within the Pakistani military establishment, as the military leadership has always been dominated by the Punjabis and Pathans. Musharraf does not belong to either of these two ethnic groups.

Pakistan's internal problems of ethnic divisions, religious violence, economic conflicts, and a strong fundamentalist presence have worsened. Political commentators argued long back that unless Pakistan society is reshaped from top to bottom, there is very little likelihood of any fundamental change (Ali 1983:195). The civilian population has hardly any legitimate redress to their grievances.

Pakistan is the United States' strategic partner in South Asia: an ally in the fight against the Taliban remnants and fundamentalists in Afghanistan, and a guarantor of US access to a sea route for Central Asian oil and gas (Nath 2004: 47). The United States' own interests have always taken priority over India's in any dispute between India and Pakistan. Virtually unqualified US support to the Pakistani military administration despite its appalling record of subverting democracy and making unauthorized sales of nuclear technology to third countries is clear evidence of that.[17] In the recent years there are indications of some small shifts in the United States' position. Recently four powerful senators issued a public statement asking the Pakistani president to stop attacks on India from the Pakistani soil by terrorists.[18]

In Pakistan, power within the state apparatus lies effectively in the hands of a military-bureaucratic oligarchy (Alavi 1983), and some eighty landed gentries control about one-tenth of the cultivated land in three of the richest provinces. In this environment, anti-Indian forces, particularly the extremists, may find water-related disputes a good rallying point for achieving their political objectives.

Domestic political events in Pakistan are relevant to India's freshwater-management policy for a number of reasons. First, Pakistan is a water-scarce country and the situation is only getting worse. The government's failure to deal with this problem effectively will make the internal political environment even more volatile. Second, Pakistan and India are competing for the same precious resource. Disputes are already arising, as both countries intend to access additional water from the Indus systems.

India and Pakistan are engaged in bilateral discussions with a view to resolving outstanding disputes. In spite of Pakistan's commitment to resolve all issues through dialogue, Pakistan's government uses every opportunity to remind people that it will not shift from its tough posture on the Kashmir dispute. And in spite of the new bonhomie between these two countries, the government lets no opportunity go by without making a public announcement of its intention to thwart India's attempt to get a permanent seat at the UN Security Council.

In 2006 the International Crisis Group confirmed that the Pakistan military tended to retain its central role along with many jihadi groups (albeit with a new identity) in the distribution of humanitarian aid in the earthquake affected region. It suggested that the international humanitarian organizations shift their approach "from an embedded relationship with the military to an effective partnership with elected officials, and credible and moderate civil society organizations."[19] The Group argued that by accepting a major role for banned jihadi groups in humanitarian relief efforts, the government's policies were helping Islamic radicals to bolster their presence in the earthquake-effected areas of the North West Frontier province and Pakistan-administered Kashmir. Both the Afghan government and allied commanders in Afghanistan have publicly accused the Taliban insurgencies of maintaining a strong base in the Pakistan city Quetta, from where they launch their attack on government positions in the Kandahar Region in southern Afghanistan, although the Pakistan government routinely refutes such charges (Baker 2007).

The internal situation instead of improving appears to have deteriorated further. In 2007 Crisis Group continued with its warning that the political situation in Pakistan was deteriorating.[20] A number of indicators confirm this. Firstly, a female cabinet minister was assassinated by fundamentalists for not wearing "proper Islamic dress"; secondly, the Chief Justice of the Pakistan Supreme Court was unceremoniously suspended by the President on misbehaviour grounds and at this the entire legal fraternity actively protested, which took an ugly turn when they confronted the authorities; thirdly, the BBC reported that in mosques across Pakistan the call was going out for men to join a jihad against NATO forces in Afghanistan, and many never returned.

Given past history and shifting political alliances—nationally and internationally—and the absence of a strong international protocol to resolve water-sharing disputes between riparian countries within a regional framework, there exists every possibility that any water-related dispute between India and Pakistan will be allowed to ferment for the internal political reasons, particularly when the forecast is that Quetta (alleged to be the base of the Taliban loyalists) will be left with no freshwater within a decade or two.

Notes

1. *Forbes* (2002) "The Global 500"; http://www.forbes.com/global/2002/0722/global.html (October 30, 2004).

2. Income distribution figures relate to 2001 and poverty level figures refer to 1995/1996 FY estimates.

3. Statement to that effect by the Government Planning Secretary on February 26, 2004, during the release of the "Poverty Map" by the Bangladesh Bureau of Statistics and the UN World Food Program, Dhaka; http://story.news.yahoo.com/news?tmpl=story ; (February 27, 2004).

4. *Weekly Times* (2002), October 21.

5. *Hindu* (2004), http://us.f502.mail.Yahoo.com October11.

6. *Daily Star* 4/260 and 4/283.

7. *Asian Times Online* (March 11, 2004) and Yahoo news (October 26, 2004). Also, Bangladesh Finance Minister Mr. Saifur Rahman even claimed on March 22,

2006, in New Delhi that there was no illegal migration, as the economic situation in Bangladesh in some areas was much better than in India.

8. *Daily Star* 4/116.

9. Independent assessment was undertaken by "Probe International." Comments by Pearce (1992), Rich (1994), McCully (1996), and others confirm the enormous power of the dam-industry and related professional bodies to preserve their self-interest.

10. *NY Times* (2004), April 9.

11. Yale Global (2004), July 24. Also see Shih, Chih-yu (1990: 123).

12. Members are: Republic of Kazakhstan, People's Republic of China, Kyrgyz Republic, Russian Federation, Republic of Tajikistan, and Republic of Uzbekistan.

13. *NY Times* (2004), December 8 and December 31.

14. Integrated Regional Information Network (2007), Pakistan: Focus on Water Crisis—Quoting an early-1990s report. The humanitarian news and analysis service of the UN Office for the Coordination of Humanitarian Affairs.

15. Also see K. Bahadur (1986).

16. *NY Times* (2004), November 8.

17. Also see Newberg (2006).

18. Senators Biden, Kerry, Leahy, and Lincoln on March 25, 2007, "US senators ask Pervez to stop attacks on India"; *Statesman*, March 26.

19. "Pakistan Political Impact of the Earthquake: A Report by the International Crisis Group"; March 2006.

20. *Crisis Watch* No. 44, April 1.

References

Abbasi, S. E. (1991), Environment impact of water resource projects. New Delhi: Discovery Publishing House.

Abraham, T. K. (2005), *Village beats tsunami with tree power*. http://in.news.yahoo.com/050128/137/2j9vx.html (November 16, 2007).

Adam, D. (2007, February 20), Climate change: Scientists warn it may be too late to save the ice caps. *The Guardian*.

Adams, J. R. and D. Frantz (1992), A Full Service Bank: How BCCI stole billions; Around the world. New York: Pocket Books.

Adams, P. (1988), Dam busters. *New Internationalist*, Issue 183.

Agarwal, A., S. Narayan and I. Khurana (Eds.) (2001), Making water everybody's business: Practice and policy of water harvesting. New Delhi: Centre for Science and Environment.

Ahamed, E. (1998), *The military and democracy in Bangladesh*. In R. J. May and V. V. Selochan (Eds.), The military and democracy in Asia and the Pacific. Canberra: ANU E Press.

Alavi, H. (1983), *Class and State*. In H. N. Gardezi and J. Rashid (Eds.), Pakistan: The root of dictatorship; The political economy of a praetorian state. London: Zed Press.

Alexis, L. (1984), *Sri Lanka's Mahaweli Ganga Project: The damnation of Paradise*. In E. Goldsmith and N. Hildyard (Eds.), Social and environmental impacts of large dams, vol. ii. Camelford, Cornwall, U.K.: Wadebridge Ecological Centre.

Ali, C. R. (1942), What does the Pakistan national movement stand for? Cambridge: Cambridge University Press.

Ali, Md., G. E. Radosevich and A. A. Khan (Eds.) (1987), Water resources policy for Asia. Rotterdam: A. A. Balkema.

Ali, Tariq (1983), Can Pakistan survive? The death of a state. London: Penguin Books.
Alter, S. (2001), Sacred waters: A pilgrimage up the Ganges River to the source of Hindu culture. New York: Harcourt.
Altman, D. (2006, November 17), Managing globalization: The trouble with water. *The International Herald Tribune.*
Appan, A. (2001), *Glimmers of Hope.* In Agarwal, A., S. Narayan and I. Khurana (Eds.), Making water everybody's business: Practice and policy of water harvesting. New Delhi: Centre for Science and Environment.
Arora, G. K. (2004), Globalisation and reorganising Indian states: Retrospect and prospect. New Delhi: Bookwell.
Arthur, Rt. Hon. O. (2001), *Opening statement at the small state in the global economy conference.* In D. Peretz, R. Faruqi and E. J. Kisanga (Eds.), Small states in the global economy. London: Commonwealth Secretariat.
Asian Development Bank (2003), Asian development outlook 2003. Manila: Oxford University Press.
Asthana, R. (1996), Involuntary resettlement: Survey of international experience. *Economic and Political Weekly*, 31/24.
Athukorala, C. P. and S. Jayasuriya (1994), Macroeconomic policies: Crises and growth in Sri Lanka 1960 to 1990. Washington D.C.: World Bank.
Austin, C. (2004), Water, wit and wisdom: The search for solutions to the water crisis. Lilydale (Victoria): WaterRight.
Ayibotele, M. B. (1992), The world's water: Assessing the resource. Conference on Water and the Environment: Development Issues for the 21st Century. Dublin, Ireland.
Bahadur, K. (1986), *Military and politics in Pakistan.* In U. Phadnis, S. D Muni and K. Bahadur (Eds.) (1986), Domestic conflicts in South Asia. New Delhi: South Asian Publishers.
Baker, A. (2007, March 27), Truth about Talibanism. *Time Magazine.*
Bakshi, J. (2004), Russia-China relations: Relevance for India. New Delhi: Shipra Publications.
Balfour, The Earl of (1936), "Foreword" in W. Bagehot, English constitution. Oxford: Oxford University Press.
Bandyopadhyay, J. (1988), The ecology of drought and water scarcity. *Ecologist* 18/2–3.
Bandyopadhyay, J. (1989), Riskful confusion of drought and man-induced water scarcity. *AMBIO* 18/5.
Bandyopadhyay, J. (1995), Sustainability of big dams in Himalayas. *Economic and Political Weekly*, 30/38.
Bandyopadhyay, J. and S. Perveen (2004), Interlinking of rivers in India: Assessing the justifications. *Economic and Political Weekly*, 39/50.
Bandyopadhyay, P.K. (2004), Tulsi leaves and the Ganges water: Slogan of the first Sepoy Mutiny at Barrakpore 1825. Kolkata: K. P. Bagchi.

Banerjee, P. (2005), *Resisting erasure: Women IDPs in South Asia*. In P. Banerjee, S. Basu Ray Chaudhury and S. K. Das (Eds.), The relevance of the UN's guiding principles. New Delhi: Sage.

Barber, B. (1995), Jihad vs. McWorld. New York, Random House.

Barber, M. and G. Ryder (Eds.), (1993), Damming the Three Gorges (2nd edition). London: Probe International and Earthscan.

Barbier, E. B. (2005), Natural resources and economic development. New York: Cambridge University Press.

Bardach, J. E. (1989), Global warming and the coastal zone: Some effects on sites and activities. *Climatic Change* 15/ 1–2.

Bardhan, P. (1999), *The economist's approach to agrarian structure*. In R. C. Guha and J. P. Pary (Eds.), Institutions and inequalities in honour of André Bétteille. New Delhi: Oxford University Press.

Barman, D. C. (2004), *Forced migration in South Asia: A study of Bangladesh*. In O. P. (Ed.), Forced migration in the South Asian region. Kolkata: Jadavpur University.

Barot, N. and S. Mehta (2001), *Women and water harvesting*. In A. Agarwal, A., S. Narayan and I. Khurana (Eds.), Making water everybody's business: Practice and policy of water harvesting. New Delhi: Centre for Science and Environment.

Barraclough, S. L. (2001), Toward integrated and sustainable development? Geneva: UN Research Institute for Social Development.

Barraclough, S. L. and K. Ghimrie (1995), Forests and livelihoods. The social dynamics of deforestation in developing countries. New York: St. Martin's Press.

Basu, K. (1997, November 17), Paying for education—Leave higher learning to the market and primary education to the states. *India Today*.

Batley, R. and G. Larbi (2004), The changing role of government: The reform of public services in developing countries. New York: Palgrave Macmillan.

Baumol, W. J. and W. E. Oates (1975), Theory of environmental policy: Externalities, public outlays and the quality of life. New Jersey: Prentice Hall.

Baxter, C. (1989), Struggle for development in Bangladesh. *Current History*, 88/542.

Behura, N. K. and P. K. Nayak (1993), *Involuntary displacement and changing frontiers of kinship: A study of resettlement in Orissa*. In M. M. Cernea and S. E. Guggenheim (Eds.), Anthropological approaches to resettlement: Policy, practice and theory. Boulder: Westview Press.

Bell, J. (1994), Hydrodollars in the Himalaya. *Ecologist*, 24/3.

Bellman, E. (2004, August 23), India waters, its grassroots. *Wall Street Journal*.

Bhagwati, J. (1996), K. R. Narayan Oration. Canberra: Australian National University.

Bhambri, C. P. (1986), *Indian political system: Nature of contradictions*. In U. Phadnis, S. D Muni and K. Bahadur (Eds.), Domestic conflicts in South Asia. New Delhi: South Asian Publishers.

Bhaumik, S. (2005), *India's North-East: Nobody's people in no-man's land*. In P. Banerjee, S. Basu Ray Chaudhury and S. K. Das (Eds.), The relevance of the UN's guiding principles. New Delhi: Sage.

Bhumbla, B. R. (1984), *Canal irrigation and its impact on crop production and environment*. In E. Goldsmith and N. Hildyard (Eds.), Social and environmental impacts of large dams, vol. i. Camelford, Cornwall, U.K.: Wadebridge Ecological Centre.

Bhutto, Z. A. (1969), Myth of independence. Lahore: Oxford University Press.

Biswas, A. K. (2001), *Management of international rivers: opportunities and constraints*. In A. K. Biswas and J. I. Uitto (Eds.), Sustainable development of the Ganges-Brahmaputra-Meghna Basin. Tokyo: UN University Press.

Blatter, J., H. Ingram and P. M. Doughman (2001), *Emerging approaches to comprehend changing global contexts*. In J. Blatter and H. Ingram (Eds.), Reflection on water—New approaches to transboundary conflicts and co-operation. Cambridge: MIT Press.

Blatter, J., H. Ingram and S. L. Levesque (2001), *Expanding perspectives on transboundary waters*. In J. Blatter and H. Ingram (Eds.), Reflection on water—New approaches to transboundary conflicts and co-operation. Cambridge: MIT Press.

Blinkhorn, T. A. and W. T. Smith (1995), *India's Narmada: River of hope*. In W. F. Fisher (Ed.), Towards sustainable development? Struggling over India's Narmada River. New York: M. E. Sharp.

Blomquist, W. and H. M. Ingram (2003), Boundaries seen and unseen: Resolving transboundary groundwater problem; *Water International*, 28/2.

Bogardi, J. J. and A. Szollosi-Nagi (2004), *Towards the water policies for the 21st century: A review after the world summit on sustainable development in Johannesburg*. In E. Cabrera and R. Cobacho (Eds.), Challenges of the new water policies for the XXI century. Lisse: Swets and Zeitlinger.

Bolinget, W. (2004), A historic testament to the resolute Cordillera peoples' struggle. *Bu-la-lat*, 4/12. Quezon City.

Borenstein, S. (2007), Warming "likely" man-made, unstoppable. http://www.washingtonpost.com/wp-dyn/content/article/2007/02/01/AR2007020100395_pf.html (November 12, 2007).

Borole, D. V., S. K. Gupta, S. Krishnaswami, P. S. Datta and B. I. Desai. (1978), Uranium isotopic investigations and radiocarbon measurements of river-groundwater systems, Sabarmati Basin, Gujrat'; *Isotope Hydrology* vol. i.

Bowker, R. (1996), Beyond peace: The search for security in the Middle East. London: Lynne Rienner Publishers.

Boyle, A. (1996), *Role of international human rights law in the protection of the environment*. In A. E. Boyle and M. R. Anderson (Eds.), Human rights: Approaches to environmental protection. Oxford: Clarendon Press.

Bradnock, R. W. (1992), *Changing geography of the states of the South Asian periphery*. In G. P. Chapman and K. M. Baker (Eds.), Changing geography of Asia. London: Routledge.

Brammer, H. (1990), Floods in Bangladesh 1. *Geographical Journal*, 156/1.

Brammer, H. (1990a), Floods in Bangladesh 2. *Geographical Journal*, 156/2.

Bray, J., L. Lunde and S. Mansood Murshed (2003), *Nepal: Economic drivers of the Maoist insurgency*. In K. Ballentine and J. Sherman (Eds.), The political economy of armed conflict: Beyond greed and grievance. Boulder: Lynne Rienner Publishers.

Breitmeier, H. and V. Rittberger (2000), *Environmental NGOs in an emerging global civil society.* In P. S. Chasek (Ed.), The global environment in the twenty-first century: Prospects for international cooperation. Tokyo: UN University Press.

Breman, J. (1999), *Ghettoization and communal politics: The dynamics of inclusion and exclusion in the Hinduvta landscape.* In R. C. Guha and J. P. Pary (Eds.), Institutions and inequalities in honour of André Bétteille. New Delhi: Oxford University Press.

Briscoe, J. and R. P. S. Malik (2006), India's water economy: Bracing for a turbulent future. New Delhi: World Bank and Oxford University Press.

Brown, B. E. and J. C. Ogden (1993), Coral bleaching. *Scientific American,* 268/1.

Brown, L. R., C. Flavin and S. Postel (1991), Saving the planet. New York: W. W. Norton & Company.

Brown, M. E. (1997), *Impact of government policies on ethnic relations.* In M. E. Brown and S. Ganguly (Eds.), Government policies and ethnic relations in Asia and the Pacific. Cambridge, Mass.: MIT Press.

Browne, J. (2004), Beyond Kyoto. *Foreign Affairs,* July–August.

Burke, J. (2002), *Land and Water Systems: Managing the Hydrological Risk.* In K. Prasad (Ed.), Water resources and sustainable development. New Delhi: Shipra Publications.

Cano, G. J. (1989), Development of the law of international water resources and the work of the International Law Commission. *Water International,* 14/4.

Caponera, D. A. (1987), *International water resources law in the Indus basin.* In Ali, Md., G. E. Radosevich and A. A. Khan (Eds.), Water resources policy for Asia. Rotterdam: A. A. Balkema.

Carlton, J. (2004, August 23), Businesses' thirst for water is unsated. *Wall Street Journal.*

Catherwood, C. (2003), Christians, Muslims and Islamic rage. Grand Rapids, Mich.: Zondervan.

Centre for Science and Environment (1991), Floods, flood plains and environmental myths: Third citizens' report on the state of India's environment. New Delhi.

Cernea, M. M. (1986), Involuntary resettlement in bank-assisted projects: A review of the applications of bank policies and procedures in FY 79–85 projects. Washington, D.C.: World Bank.

Cernea, M. M. (1990), *Poverty risks from population displacement.* In Water resource development. Cambridge, Massachusetts, Harvard Institute for International Development.

Cernea, M. M. (1994), *Sociologist's approach to sustainable development.* In I. Serageldin and A. Steer (Eds.), Making development sustainable: From concepts to action. Washington D.C.: World Bank.

Cernea, M. M. (1996), Public policy responses to development-induced population displacements. *Economic and Political Weekly,* 31/24.

Chakma, M. K. (2004), *Recurring displacement is made: A case of indigenous people of Chittagong Hill Tracts, Bangladesh.* In O. P. Misra (Ed.), Forced migration in the South Asian region. Kolkata: Jadavpur University.

Chakraborty, R. (2004), Sharing of water among India and its neighours in the 21st Century: War or peace? *Water International*, 29/2.
Chakravarty, C. (2004, November 17), Water, water everywhere, not a drop to drink. www.theworldjournal.com/2003/waterdrop.htm.
Chambers, R. (1988), Managing canal irrigation: Practical analysis from South Asia. Cambridge: Cambridge University Press.
Chamie, J. (2004), Coping with world population boom and burst—Part I. Yale Centre for Study of Globalization. http://yaleglobal.yale.edu/display.article?id=4389 (November 12, 2007).
Chandler, D. (2004), Construction of global civil society: Morality and power in international relations. New York: Palgrave Macmillan.
Chandler, W. (1984), Myth of the TVA; Ballinger. Cambridge, Mass.
Chandler, W. (1986), *Tellico and Columbia dams: Stewardship and development*. In E. Goldsmith and N. Hildyard (Eds.), Social and environmental impacts of large dams, vol. ii. Camelford, Cornwall, U.K.: Wadebridge Ecological Centre.
Chandra, K. (2004), *Elite incorporation in multiethnic societies*. In A. Varshney (Ed.), India and the politics of developing countries: Essays in memory of Myron Weiner. New Delhi: Sage.
Chandrakanth, M. G. and J. Romm (1990), Groundwater depletion in India— Institutional management regimes. *Natural Resource Journal*, Summer 1990.
Chapman, G. P. (1992), *Changes in the South Asian core: Patterns of growth and stagnation in India*. In G. P. Chapman and K. M. Baker (Eds.), Changing geography of Asia. London: Routledge.
Chapman, G. P. (1995), *The Ganges and Brahmaputra Basins*. In G. P. Chapman and M. Thompson (Eds.), Water and the quest for sustainable development in the Ganges Valley. New York: Mansell Publishing.
Chapman, G. P. (1995a), Environmental myth as international politics: The problem of the Bengal Delta. In G. P. Chapman and M. Thompson (Eds.), Water and the quest for sustainable development in the Ganges Valley. New York: Mansell Publishing.
Chapman, G. P. (2000), Geopolitics of South Asia: From early empires to India, Pakistan and Bangladesh. Aldershot: Ashgate.
Chapman, G. P. and M. Thompson (Eds.) (1995), Water and the quest for sustainable development in the Ganges valley. New York: Mansell Publishing.
Charbeneau, R. J. (1982), Groundwater resources of the Texas Rio Grande Basin. *Natural Resource Journal*, 22/4.
Chari, P. R. (2003), *Refugees, migrants and internally displaced persons in South Asia: An overview*. In P. R. Chari, M. Joseph and S. Chandran (Eds.), Missing boundaries: Refugees, migrants, stateless and internally displaced persons in South Asia. New Delhi: Manohar Publishers.
Chatterjee, D. (2004), *Dalits in India as development's victims: The twice dammed*. In O. P. Mishra (Ed.), Forced migration in the South Asian region. Kolkata: Jadavpur University.

Chaudhuri, K. and S. Dasgupta (2006), Political determinants of fiscal policies in the states of India: An empirical investigation. *Journal of Development Studies*, 42/4.

Chellaney, B. (2004), Drawing the line with China. *Japan Times* online. www.japantimes.com/cgi-bin/geted.pl5?eo20040730bc.htm (7/31, 2004).

Chellaney, B. (2005), India-China rivalry sharpens; *Japan Times* online. www.japantimes.com/cgi-bin/makeprfy.pl5?eo20050402bc.htm (4/6, 2005).

Chinnammai, S. (2006), *Region-wise interlinking of rivers—An overview*. In S. R. Singh and M. P. Shrivastava (Eds.), River interlinking in India: The dream and reality. New Delhi: Deep and Deep Publications.

Chopra, K. and B. Goldar (2002), *Sustainable use of water in India: Implications of alternative scenarios for 2020*. In K. Prasad (Ed.), Water resources and sustainable development. New Delhi: Shipra Publications.

Chopra, P. N. (1986), Multi-level planning in India. New Delhi: Intellectual Publishing.

Chorus, I. and J. Bartram (Eds.) (1999), Toxic cyanobacteria in water. London and New York: F & FN Spon.

Chung, Chien-peng (2004), Domestic politics, international bargaining and China's territorial disputes. London: RoutledgeCurzon.

Clarke, R. (1991), Water: The international crisis. Cambridge, Mass.: MIT Press.

Cohen, R. and F. M. Deng (1998), Masses in light: The global crisis of internal displacement. Washington D.C.: Brookings Institute Press.

Colomer, J. M. (2001), Political institutions: Democracy and social choice. Oxford: Oxford University Press.

Connelly, P. and R. Perlman (1975), Politics of society: Resource conflicts in international relations. London: Royal Institute of International Affairs and Oxford University Press.

Connor, S. (2007, January 29), Global warming: It's much worse than you thought. *Independent*.

Corell, E. and A. Swain (1995), *India: The domestic and international politics of water scarcity*. In L. Ohlsson (Ed.), Hydropolitics: Conflicts over water as a development constraint. Dhaka: University Press.

Cornia, G. A. and J. Court (2001), Inequality, growth and poverty in the era of liberalization. Helsinki: UN University Press.

Crow, B. (1995), Sharing the Ganges: Politics of technology of river development. New Delhi: Sage.

Cumberland, J. H. and H. W. Herzog Jr. (1977), Effects of economic development on water resources (Technical Paper No. 40). University of Maryland Water Resources Research Centre.

D'Souza, R. (2006), Interstate disputes over Krishna waters: Law, science and imperialism. New Delhi: Orient Longman Pvt. Ltd.

Dahal, D. R. (2001), Civil society in Nepal: Opening the ground for questions. Kathmandu: Centre for Development and Governance.

Dallmayr, F. and G. N. Devy (Eds.) (1998), Between tradition and modernity. Walnut Creek, Calif.: AltaMira.

Darwin, R. (2001), Climate change and food security. Economic Research Service, US Department of Agriculture, June 2001.

Dasgupta, B. (2000), *International institution for global trade: The case for South Asian free trade association.* In D. K. Dutta (Ed.), Economic liberalization and institutional reform in South Asia: Recent experiences and future prospects. New Delhi: Atlantis Publishers.

Davies, Bill and P. Dry (2003, November 29), *Drought Crops* (Interview with the Australian Broadcasting Commission). www.abc.net.au/rn/science/ss/stories/s990372.htm (December 2, 2003).

De Villiers, M. (1999), Water. Toronto: Stoddard Publishing.

Dellapenna, J. W. (1999), Custom-built solutions for international disputes. www.unesco.org/courier/1999_02/uk/dossier/txt41.htm (March 10, 2004).

Deshpande, J. V. (1997), Behind Dalit anger. *Economic and Political Weekly*, 32/33–34.

Devine, R. S. (1995), Trouble with dams. *Atlantic Monthly*, 276/2.

Dhagamwar, V., E. Ganguly-Thukral and M. Singh (1995), *The Sardar Sarovar project: A study in sustainable development?* In W. F. Fisher (Ed.), Toward sustainable development? Struggling over India's Narmada river. New York: M. E. Sharp.

Dhakal, S. (2003), Global warming threatens lake bursts in Nepal, in *One World South Asia*. http://www.globalpolicy.org/socecon/envronmt/2003/1209nepal.htm (November 12, 2007).

Dhar Chakraborti, R. (2004), The greying of India: Population ageing in the context of Asia. New Delhi: Sage.

Dharmadhikary, S. (1995), *Hydropower at Sardar Sarovar: Is it necessary, justified and affordable?* In W. F. Fisher. (Ed.), Toward sustainable development? Struggling over India's Narmada river. New York: M. E. Sharp.

Dias, A. (2002), *Development-induced displacement and its impact.* In S. Tharakan (Ed.), The nowhere people: Responses to internally displaced persons. Bangalore: Books of Change.

Dickens, G. R. (2004), Hydrocarbon-driven warning. *Nature* 429/6991.

Dickinson, R. E. (1989), Uncertainties of estimates of climate change: A review. *Climatic Change* 15/1–2.

Dinar, A., T. K. Balkrishnan and J. Wambia (1998), Political economy and political risks of institutional reforms. Water sector policy research working paper. Washington, D.C.: World Bank.

Djurfeldt, G. and S. Lindberg (1975), Behind poverty: Social formation in a Tamil village. Scandinavian Institute of Asian Studies. Lund: Curzon Press.

Dogra, B. (1986), *Indian experience with large dams.* In E. Goldsmith and N. Hildyard (Eds.), Social and environmental impacts of large dams vol. ii. Camelford, Cornwall, U.K.: Wadebridge Ecological Centre.

Dogra, B. (1992), Debate on large dams. New Delhi: Centre for Science and Environment.

Dogra, B. (1996), Land reforms to fight hunger. *Economic and Political Weekly* 31/ 40.

Donahue, M. and B. R. Johnston (Eds.) (1998), Water, culture and power in a global context. Washington D.C.: Island Press.

Doolette, J. B. and W. B. Magrath (Eds.) (1990), Watershed development in Asia: Strategies and technologies. Washington D.C.: World Bank.

Dorcey, T., A Steiner, A. M. Acreman and B. Orlando (Eds.) (1997), Large dams: Learning from the past, looking at the future. Gland: IUCN and the World Bank.

Dror, Y. (1986), Policymaking under adversity. Transaction Books, New Brunswick.

Dubash, N. K, M. Dupar, S. Kothari and T. Lissu. (2001), Watershed in global governance: An independent assessment of the World Commission on Dams. Washington D.C.: World Resources Institute.

Dubash, N. K. (2002), Tubewell capitalism: Groundwater development and agrarian change in Gujrat. New Delhi: Oxford University Press.

Duchacek, I. D. (1986), Territorial dimension of politics within, among, across nations. Boulder: Westview Press.

Dupont, A. and G. Pearman (2006), Heating up the planet: Climate change and security. Sydney: Lowy Institute.

Dutta, S. (1998), *China's emerging power and military role: Implications for South Asia*. In J. D. Pollack and R. H. Yang (Eds.), In China's shadow: Regional perspectives on Chinese foreign policy and military development. Washington D.C.: RAND.

Dyson, T, T. R. Cassen and L. Visaria (Eds.) (2004), Twenty-first century India: Population, economy, human development and the environment. New Delhi: Oxford University Press.

Easter, K. W. and K. Palarisami (1985), Tank irrigation in India: An example of property resource management. Proceedings of the conference on common property. Washington D.C.: International Development Office of International Affairs, National Research Council.

Eckstein, Y. and G. E. Eckstein (2003), Groundwater resources and international law in the Middle East peace process. *Water International*, 28/2.

Economic and Social Council for Asia Pacific (1992), State of the environment in Asia and the Pacific 1990. Bangkok: ESCAP.

Economy, E. (2007, January 26), China's missile message. *The Washington Post*.

Elkin, S. L. (1996), *The constitution of good society: The case of the commercial republic*. In K. E. Soltan and S. L. Elkin (Eds.), Constitution of good societies. University Park: Pennsylvania University Press.

Elmusa, S. S. (1997), Economics, politics, law and Palestinian-Israel water resources. Washington, D.C.: Institute for Palestine Studies.

Emery, F. E. and E. L. Trist (1972), Toward a social ecology. New York: Plenum Press.

Emmons, J. (1990), Weighing hydroelectric options in Nepal: Some say, "Keep it small." *Water International*, 15/2.

Emmott, B. (2003), 20:21 VISION—Lessons of the 20th Century for the 21st. London: Allen Lane, an imprint of Penguin Books.

Endersbee, L. A. (2005), A voyage to discovery: A history of ideas about the earth. Frankston: Lance: Endersbee.

Fair, C. C. and S. Ganguly (2007, February 7), Bangladesh on the brink. *Wall Street Journal*.

Falkenmark, M. (1989), Vulnerability generated by water scarcity. AMBIO, 18/6.

Falkenmark, M. (1998), *Meeting water requirements of an expanding world population*. In D. J. Greenland, P. J. Gregory and P. H. Nye (Eds.), Land resources: On the edge of the Malthusian precipice? London: CABI in association with Royal Society.

Falkenmark, M. and G. Lindh (1975), Water for a starving world. Boulder: Westview Press.

Falkenmark, M., J. Lundgvist, S. Postel, J. Rockstrom, D. Seckles, H. Shuval, et al. (1998), Water scarcity as a key factor behind global food insecurity. Round table discussion. AMBIO, 27/2.

Falkenmark, M., L. da Cunha, and L. David (1987), New water management strategies needed for the 21st century. *Water International*, 12/3.

Ferdinand, F. (2002), *China and Central Asia*. In K. E. Brodsgaard and B. Heurlin (Eds.), China's place in global geopolitics. London: RoutledgeCurzon.

Fernandes, W. (1991), Power and powerlessness: Development projects and displacement of tribals. *Social Action*, 41.

Fernandes, W. and V. Paranjpye (Eds.) (1997), Rehabilitation policy and law in India: A right to livelihood. New Delhi, Indian Social Institute.

Finger, M. and J. Allouche (2002), Water privatisation (Trans-national corporations and the re-regulation of the water industry). London: Spon Press.

Fisher, D. R. (2004), National governance and the global climate change regime. New York: Rowman and Littlefield.

Fisher, David (2005), *Epilogue: International law on the internally displaced person*. In P. Banerjee, S. Basu Ray Chaudhury and S. K. Das (Eds.), The relevance of the UN's guiding principles. New Delhi: Sage.

Flannery, T. (2005), Weather makers. New York: Atlantic Monthly Press.

Flyvbjerg, B., N. Bruzelius and W. Rothengatter (2003), Megaprojects and risk: An anatomy of ambition. Cambridge, Cambridge University Press.

Forrest, R. (1991), Japanese aid and the environment. *Ecologist* 21/1.

Frederiksen, H. D., J. Berkoff and W. Barber (1993), Water resource management in Asia, vol. 1, Main report: World Bank technical paper No. 212, Washington D.C: World Bank.

Frey, F. W. (1993), Political context of conflict and cooperation over international river basins. *Water International*, 18/1.

Friedkin, J. F. (1987), *International water treaties: United States and Mexico*. In Ali, Md., G. E. Radosevich and A. A. Khan (Eds.), Water resources policy for Asia. Rotterdam, A. A. Balkema.

Gan, L. (2000), *Energy development and environmental NGOs: Asian Perspectives*. In P. S. Chasek (Ed.), The global environment in the twenty-first century: Prospects for international cooperation. Tokyo: UN University Press.

Ganguly-Thukral, E. (1996), Development, displacement and rehabilitation: Locating gender; *Economic and Political Weekly*, 31/24.

Gelbspan, R. (1997), The heat is on: The high stakes battle over earth's threatened climate. New York: Addison-Wesley Publishing Co. Inc.

George, S. J. (2001), *Politics of ethnicity and autonomy: The Jharkhand movement*. In S. J. George (Ed.), Intra and inter-state conflicts in South Asia. New Delhi: South Asian Publishers.

Ghai, D. (1997), *Economic globalization, institutional changes and human society*. In S. Lindberg and A. Sverrisson (Eds.), Social movement in development: The challenge of globalization and democratization. New York: St. Martin's Press.

Ghassemi, F., A. J. Jakeman and H. A. Nix (1995), Salinisation of land and water resources: Human causes, extent, management and case studies. Sydney: University of NSW Press.

Ghosh, R. and T. S. Bathija (2002), *Review of constitutional provisions governing water resources of India*. In K. Prasad (Ed.), Water resources and sustainable development. New Delhi: Shipra Publications.

Gijsberts, M., L. Hagendoorn and P. Scheepers (Eds.) (2003), Nationalism and exclusion of migrants—Cross-national comparisons. Aldershot, Ashgate.

Gilmartin, D. (1994), Scientific empire and imperial science: Colonialism and the Indus basin. *Journal of Asian Studies*, 53/4.

Gleick, P. H. (1996), Basic water requirements for human activities: Meeting basic needs; *Water International*, 21/2.

Gleick, P.H.(1998), World's water: The biennial report on freshwater resources, 1998–99: Table 16. California, Pacific Inst.

Gleick, P. H.(1999), Human right to water. *Water Policy*, 1/5.

Gleick, P. H. (2000), World's water 2000–2001: The biennial report on freshwater resources. Washington D.C.: Island Press.

Gleick, P. H. (2000a), *A soft path: Conservation, efficiency, and easing conflicts over water*. In P. S. Chasek (Ed.), Global environment in the twenty-first century: Prospects for international cooperation. Tokyo: UN University Press.

Gleick, P. H. (2003), *A soft path: Conservation, efficiency and easing conflicts over water*. In B. McDonald and D. Jehl (Eds.), Whose water is it? The unquenchable thirst of a water-hungry world; Washington D.C.: National Geographic Society.

Gleick, P. H. (2003a), Water conflict chronology. www.worldwater.org/conflict chronology.pdf (March 29, 2004).

Gleick, P. H. (2004), US per-capita water use falls to 1950 levels. http://www.pacinst.org/press_center/usgs/ (November 12, 2007).

Gleick, P. H. (2007), www.pacinst.org/publications/testimony (March 29, 2007).

Goel, R. S. (2003), *Big versus small dams controversy: A critical overview of socio-economic and environmental concerns in Indian context*. In K. Prasad (Ed.), Water resources and sustainable development. New Delhi: Shipra Publications.

Goldsmith, E. (1988), Aid: Enlightened self-interest or gun-boat politics. *Ecologist*, 18/ 2–3.

Goldsmith, E. and N. Hildyard (1984–92), The social and environmental effects of large dams, 3 vols. Camelford, Cornwall, U.K.: Wadebridge Ecological Centre.

Gopalkrishnan, M. (2002), *Role of large dams in India*. In K. Prasad (Ed.), Water resources and sustainable development. New Delhi: Shipra Publications.

Gore, A. (2006), Inconvenient Truth. Emmaus: Rodale.

Gossman, P. (2002), Kashmir and international law: How war crimes fuel the conflict. http://www.crimesofwar.org/onnews/news-kashmir.html (November 12, 2007).

Gottlieb, R. (1988), A life of its own: The politics and power of water. Javanovich: Harcourt Brace.

Gounder, R. and V. Xayavong (2001), Globalization and the island economies of South Pacific. Helsinki: UN University.

Government of India (2004), A reference manual. Ministry of I&B; New Delhi.

Goyal, P. (2002, January 10), Food security in India. *The Hindu*.

Goyal, S. (1996), Economic perspectives on resettlement and rehabilitation. *Economic and Political Weekly*, 31/24.

Grynberg, R. (1998), Rules of origin: Issues in Pacific Island development. Canberra: Australian National University.

Guggenheim, S. E. and M. M. Cernea (Eds.) (1993), Anthropological approaches to resettlement: Policy, practice and theory. Boulder: Westview Press.

Guha, R. C. (2000), Environmentalism: A global history. New Delhi: Oxford University Press.

Gulati, L. and S. I. Rajan (1999), Added years: elderly in India and Kerala. *Economic and Political Weekly*, 34/44.

Gunaratna, M. H. (1988), For a sovereign state. Ratamala: Sarvodaya.

Gupta, D. (1999), *Civil society or the state: What happened to citizenship?* In R. C. Guha and J. Pary (Eds.), Institutions and inequalities in honour of André Bétteille. New Delhi: Oxford University Press.

Gupta, H. K. (1992), Reservoir-induced earthquakes. Amsterdam: Elsevier.

Hallengren, A. (Ed.) (2004), Nobel laureates—In search of identity and integrity—Voices of different cultures. Singapore: World Scientific Publishing.

Hanchett, S., J. Akhter and K. R. Akbar (1998), *Gender and society in Bangladesh's flood action plan*. In M. Donahue and B. R. Johnston (Eds.), Water, culture and power in a global context. Washington D.C.: Island Press.

Hancock, G. (1989), Lords of poverty. London: Macmillan.

Hansen, J, L. Nazarenko, R. Ruedy, M. Sato, J. Wills, A. D. Genio et al. (2005), Earth's energy imbalance: Confirmations and implications. *Science*, 308/5722. www.scienceexpress.org.

Hansen, J. (2006), Planet in Peril—Part I: Global warming, arctic ice melt and rising oceans will shrink nations and change world maps. http://yaleglobal.yale.edu/display.article?id=8305 (January 9, 2008).

Harries, O. (2005), Morality and foreign policy. Occasional paper No.14. Sydney: Centre for Independent Studies.

Harshe, R. (2001), *Understanding conflicts in South Asia*. In S. J. George (Ed.), Intra and inter-state conflicts in South Asia. New Delhi: South Asian Publishers.
Herz, B., K. Subbarao, M. Habib, and L. Raney. (1991), Letting girls learn: Promising approaches in primary and secondary education. World Bank discussion paper no. 33. Washington D.C.
Hillel, D. (1987), Efficient use of water in irrigation. World Bank technical paper no. 64. Washington, D.C.
Hillel, D. (1994), Rivers of Eden: The struggle for water and the quest for peace in the Middle East. New York: Oxford University Press.
Hillier, C. (1988), Mangrove wastelands. *Ecologist*, 18/ 2–3.
Hinton, W. (1969), PL, *Progressive Labor*, 6/6.
Hira, A, and T. Parfitt (2004), Development projects for a new millennium. London: Praeger.
Hoge, J. F., Jr. (2004), Global power shift in the making. *Foreign Affairs*, July–August.
Homer-Dixon, T. (1996), Two Indias. *Maclean's*, 109/4.
Hooper, J. (1995, June 4), Drain on Spain. *Guardian*.
Huang, Y. and D. L. Yang (2004), *Population control and state coercion in China*. In B. J. Naughton and D. L. Yang (Eds.), Holding China together: Diversity and national integration in the post-Deng era. New York: Cambridge University Press.
Hundley, N., Jr. (1966), Dividing the waters: A century of controversy between the United States and Mexico. Berkley: University of California Press.
Hunter, J. M, L. Rey, K. Y. Chu, E. O. Adekolu-John and K. E. Mott (1993), Parasitic diseases in water resource development: The need for intersectoral negotiation. Geneva: World Health Organization.
Hussain, W. (2003), *Bangladeshi migrants in India: Towards a practical solution—A view from the North-Eastern Frontier*. In P. R. Chari, M. Joseph and S. Chandran (Eds.), Missing boundaries: Refugees, migrants, stateless and internally displaced persons in South Asia. New Delhi: Manohar Publishers.
Hussein, M. (2004), *Nationalities, ethnic processes and violence in India's North East*. In R. Samaddar (Ed.), Peace studies: An introduction to the concept, scope and themes. New Delhi: Sage.
Independent Commission on International Development Issues (1980), North-south programme for survival. Cambridge: MIT Press.
Ingram, H. (1990), Water politics: Continuity and change. Albuquerque: University of New Mexico Press.
Isaak, R. A. (2004), Globalization gap: How the rich get richer and the poor get left further behind. New Jersey: FT Prentice Hall.
Iyer, R. R. (2002), *Water conflicts: A note*. In K. Prasad (Ed.), Water resources and sustainable development. New Delhi: Shipra Publications
Iyer, R. R. (2002a), Linking of rivers: Judicial activism or error. *Economic and Political Weekly*, 37/46.
Iyer, R. R. (2003), Water perspectives, issues and concerns. New Delhi: Sage.

Iyer, R. R. (2004, July 13), Punjab Bill on River Waters. *Indian Express*.
James, L. D. (1994), Flood action: An opportunity for Bangladesh. *Water International*, 19/2.
Jeevankumar, D. (2002), *Rights of internally displaced persons*. In S. Tharakan (Ed.), The nowhere people: Responses to internally displaced persons. Bangalore: Books of Change.
Jensen, M. E., W. R. Rangeley and P. Dieleman (1986), *Irrigation trends in world agriculture*. In B. A. Stewart and D. R. Nielsen (Eds.), Irrigation of agricultural crops. Madison: American Society of Agronomy.
Jhaveri, N. (1988), Three Gorges debacle. *Ecologist*, 18/2–3.
Johansson, T. B., H. Kelly, A. K. N. Reddy, R. H. Wiliams, and L. Burnham (1998), Renewable Energy: Sources for Fuels and Electricity. Washington, D.C.: Island Press.
Karns, M. P. and K. A. Mingst (2004), International organizations: The politics and processes of global governance. Boulder: Lynne Rienner.
Kashyap, S. C. (2004), Constitutional reform: Problems, prospects and perspectives (2nd edition). New Delhi: Radha Publications.
Kathuria, V. and H. Gundimeda (2002), *Industrial pollution control: Need for flexibility*. In K. S. Parikh and R. Radhakrishnan (Eds.), India development report 2002. Mumbai: Oxford University Press.
Kayastha, R. L. (2001), *Water resources development in Nepal: A regional perspective*. In A. K. Biswas and J I. Uitto (Eds.), Sustainable development of the Ganges-Brahmaputra-Meghna Basin. Tokyo: UN University Press.
Kelly, M. (2003), Science and technology: The essential guide—Climate change. www.bbc.co.uk/worldservice/sci_tech/features/essentialguide/theme_env.shtml (November 24, 2003).
Kelly, M. (2004), Climate change, visionaries week 7, Environment, programme 3. www.bbc.co.uk/worldservice/sci_tech/features/essentialguide/download/environment/prog3.rtf (November 12, 2007).
Kennedy, P. (2006), Parliament of man—The past, present and future of the United Nations. New York: Random House.
Kerkvliet, B. J. (1990), Everyday politics in the Philippines: Class and state relations in a central Luzon village. Berkley: University of California Press.
Khadka, N. S. (1994), Politics and development in Nepal. Jaipur: Nirala Publications.
Khagram, S. (2000), *Toward democratic governance for sustainable development: Transnational civil society organizing around big dams*. In A. M. Florini (Ed.), Third force: The rise of transnational civil society. Japan Centre for International Exchange and Carnegie Endowment for International Peace: Tokyo and Washington, D.C.
Khemani, S. (2004), Political cycles in a developing economy: Effect of election in the Indian states. *Journal of Development Economics*, 73/1.
King, L. C. (1983), Wandering continents and spreading sea floors on an expanding earth. Chichester: John Wiley.

Kinnersley, D. (1988), Troubled water: Rivers, politics and pollution. London: Hilary Shipman.

Kirmani, S. S. (1990), Water, peace and conflict management: The experience of the Indus and Mekong river basins. *Water International*, 15/4.

Kiss, Alexandre-Charles (2002), *Will the necessity to protect the global environment transform the law of international relations in contemporary issues*. In D. Freestone, S. Subedi and S. Davidson (Eds.), International Law: A collection of the Josephine Onoh memorial lectures. The Hague: Kluwer International.

Kolbert, E. (2006), Field notes from a catastrophe: Man, nature, and climate change. London: Bloomsbury.

Korten, D.C. (1992), Getting to the 21st century: Voluntary action and the global agenda: New Delhi: Oxford and IBH Publishing.

Kothari, A. (1995), India's dams fail environmental appraisal. *World Rivers Review*, 10/3; August 1995.

Kothari, A. (1999), *Environmental aspects of large dams in India: Problems of planning, implementation and monitoring*. In K. Prasad and R. S. Goel (Eds.), Environmental management in hydro-electric projects: Proceedings of the national seminar in India. New Delhi: Concept Publishers.

Kothari, S. (1996), Whose nation? Displaced as victims of development. *Economic and Political Weekly*, 31/24.

Kothari, S. and W. Harcourt (2004), Violence of development. *Development*, 47/1.

Kumar, S. (1983), Pauranic lore of holy water-places. New Delhi: Munshilal Monoharlal.

Kütting, G. (2004), Globalization and the environment: Greening global political economy. New York: State University of New York.

L'vovitch, M. I. (1979), World water resources and their future. *Geojournal*, 3/5. Translated by R. L. Nace from Russian.

Lahiri-Dutt, K. (2003), People, power and rivers: Experiences from the Damodar river India. *Water Nepal*, 9–10/1–2.

Laird, T. (2002), Into Tibet: The CIA's first atomic spy and his secret expedition to Lhasa. New York: Grove Press.

Lama, M. P. (1998), *Issues in harnessing economic resources in South Asia*. In B. Ghoshal (Ed.), ASEAN and South Asia development experience. New Delhi: Sterling Publishers.

Lama, M. P. (2001), *Water Resources, environment and conflicts in South Asia*. In S. J. George (Ed.), Intra and inter-state conflicts in South Asia. New Delhi: South Asian Publishers.

Lambsdorff, J. G. (2003), How corruption affects persistent capital flows. *Economics of Governance*, 4/3.

Lavergne, M. (1986), *Seven deadly sins of Egypt's Aswan high dam*. In E. Goldsmith and N. Hildyard (Eds.), Social and environmental impacts of large dams, vol. ii. Camelford, Cornwall, U.K.: Wadebridge Ecological Centre.

Lenzen, B. and B. Foran (2001), An input-output analysis of Australian water usage. *Water Policy*, 3/4.
Leslie, J. (2005), Deep water: The epic struggle over dams, displaced people, and the environment. New York: Farrar, Straus and Giroux.
Lewis, J. P. (1995), India's political economy—Governance and reform. New Delhi: Oxford University Press.
Lipper, J. (1967), *Equitable utilization*. In A. H. Garreston (Ed.), Law of international drainage Basins. New York: Dobbs Ferry.
Lloyd-Sherlock, P. (2004), *Ageing, development and social protection: Generalisations, myths and stereotypes*. In P. Lloyd-Sherlock (Ed.), Living longer, ageing, development and social protection. New York: Zed Books.
Lora, J., M. Sancho and E. Sariano (2004), Future of desalinization as a water source. In E. Cabrera and R. Cobacho (Eds.), Challenges of new water policies for the XXI century. Lisse: A. A. Balkema.
Lowie, M. R. (1999), *Transboundary resource disputes and their resolutions*. In D. H. Deudney and R. A. Mathew (Eds.), Contested grounds: Security and conflict in the new environment politics. New York: State University of New York Press.
Macnaghten, Lord. E. (1893), Water and environment. *Economic Affairs*, 18/2.
Mahmood, K. (1987), Reservoir sedimentation: Impact, extent and mitigation. World Bank Technical Paper Number 71. Washington, D.C.: World Bank.
Mamdani, M. (1990), Social movements, social transformation and the struggle for democracy in Africa. *Council for the Development of Economic and Social Research in Africa Bulletin*. Dakar.
Mandal, D. N. (2006, April 29), Bengal shining. *Statesman Weekly*, April 29.
Manzur, A. (1973), Dispute between experts. Dhaka: Minerva.
Marden, P. (2004), Decline of politics: Governance, globalization and the public sphere. Aldershot: Ashgate.
Margolis, E. S. (2006, November 17), Beijing's Africa summit: Why China is wooing the dark continent. Dubai: *Khaleej Times*.
Martone, G. (2006), *Life with dignity: What is the minimum standard*. In A. F. Bayefsky (Ed.), Human rights and refugees, internally displaced persons and migrant workers: Essays in memory of Joan Fitzpatrick and Arthur Helton. Leiden: Martinus Nijhoff Publishers.
May, R. J. (1997), *Ethnicity and the public policy in Philippines*. In M. E. Brown and S. Ganguly (Eds.), Government policies and ethnic relations in Asia and the Pacific. Cambridge: MIT Press.
McCully, P. (1996), Silenced rivers: The ecology and politics of large dams. London: Zed Books.
Mehta, L. (2000), Water for the twenty-first century: Challenges and misconceptions. Sussex: IDS Working Paper 111.
Meinzen-Dick, R. and M. Mendoza (1996), Alternative water allocation mechanisms: Indian and international experiences. *Economic and Political Weekly*, 31/13.

Messerli, B. and Hofer, T. (1995), *Assessing the impact of anthropogenic land use change in the Himalayas*. In G. P. Chapman and M. Thompson (Eds.) (1995), Water and the quest for sustainable development in the Ganges valley. New York: Mansell Publishing.

Mikesell, R. (1992), *Project evaluation and sustainable development*. In R. Goodland, H. Daly and El Sarafy (Eds.) (1992), Population, technology and lifestyle. Washington D.C.: Island Press.

Milliman, J. D., J. M. Broadus and F. Gable (1989), Environment and economic implication of rising sea level and subsiding deltas: The Nile and Bengal examples. *AMBIO*, 18/6.

Mink, S. (1994), Poverty, population and the environment. Washington, D.C.: World Bank Discussion Paper no. 189

Misra, S.D. (1970), Rivers of India: New Delhi, National Book Trust.

Mitra, A. (2003, May 30), Knotty, or just nutty? Kolkata: *The Telegraph*.

Mitra, A. K. (1996), Irrigation sector reform. *Economic and Political Weekly*, 31/13.

Mitra, P. (2006), AIDS threatens India's prosperity. http://yaleglobal.yale.edu/display.article?id=8486 (December 13, 2006)

Moench, M. (2002), Groundwater and food security in India. In K. Prasad (Ed.), Water resources and sustainable development. New Delhi: Shipra Publications.

Mohan, R. V. R. (2003), Rural water supply in India: Trends in institutionalizing people's participation. *Water International*, 28/4.

Mohsin, A. (1997), Politics of nationalism—The case of Chittagong Hill Tracts, Bangladesh. Dhaka: University Press.

Mollinga, P. P. and A. Bolding (Eds.) (2004), Politics of irrigation reform: Contested policy formulation and implementation in Asia, Africa and Latin America. Aldershot: Ashgate.

Mollinga, P. P., R. Doraiswamy and K. Engbersen (2004), Capture and transformation: Participatory irrigation management in Andhra Pradesh. In P. P. Mollinga and A. Bolding (Eds.). Politics of irrigation reform: Contested policy formulation and implementation in Asia, Africa and Latin America. Aldershot: Ashgate.

Moreira, J. R. and A. D. Poole (1993), *Hydropower and its constraints*. In T. B. Johansson, H. Kelly, A. K. N. Reddy, R. H. Wiliams, and L. Burnham, *Renewable Energy: Sources for Fuels and Electricity*. Washington, D.C.: Island Press, 1998.

Morgenthau, H. J. (1962), Politics in the twentieth century (vol. 3): Restoration of American politics. Chicago: Chicago University Press.

Morse, B. and T. Berger (1992), Sardar Sarovar: The report of the independent commission. Ottawa: Resource Future International.

Mukarji, N. (1997), Pay commission and all that. *Economic and Political Weekly*, 32/15.

Mukherjee, S. (2001), *Water resource development and the environment in Bhutan*. In A. K. Biswas and J. I. Uitto (Eds.), Sustainable development of the Ganges-Brahmaputra-Meghna Basin. Tokyo: UN University Press.

Mukherjee, S. K. (2006, January 15), River linking risks. *The Statesman*.

Mukherji, P. N. (1986), *Social conflicts and social change: Towards a theoretical orientation*. In U. Phadnis, S. D. Muni and K. Bahadur (Eds.), Domestic conflicts in South Asia, vol.1; Political dimensions. New Delhi: South Asia Publishers.
Müller, W. C. and K. Strom (2000), Coalition governments in Western Europe. Oxford: Oxford University Press.
Munir-uz-Zaman, M. (1987), *Issues relating to the uses of international water resources and World Bank practices*. In Ali, Md., G. E. Radosevich and A. A. Khan (Eds.), Water resources policy for Asia; Rotterdam, A. A. Balkema.
Muppidi, H. (2004), Politics of the Global. Borderlines, vol. 23. Minneapolis: University of Minnesota Press.
Murphy, J. (1999), Salinity—Our silent disaster www.abc.net.au/science/slab/salinity/default.htm (December 2, 2003).
Mustard, J. F. (2002), *Early child development and the brain—The base for health, learning and behaviour throughout life*. In M. E. Young (Ed.), From early child development to human development. Washington, D.C.: World Bank.
Nair, A. R., A. S. Pendharkar, S. V. Navada and S. M. Rao (1978), *Groundwater recharge studies in Maharastra*; Isotope hydrology, vol. II. Vienna: International Atomic Energy Agency.
Nath, D.C. (2004), Intelligence imperatives for India. New Delhi: India First Foundation.
National Research Council (1993), Soil and water quality: An agenda for agriculture. Washington, D.C.: National Academy Press.
Natural Resources Defense Council (2004), Energy down the drain: The hidden costs of California's water supply. www.nrdc.org/water/conservation/edrain/execsum.asp (November 12, 2007).
Naughton, B. J. (2004), *Western development program*. In B. J. Naughton and D. L. Yang (Eds.), Holding China together: Diversity and national integration in the post-Deng era. Cambridge: Cambridge University Press.
Nayar, K. (1975), Distant neighours: A tale of the subcontinent. Delhi: Vikash.
Newberg, P. R. (2006), Will Guns Fall Silent in South Asia? http://yaleglobal.yale.edu/display.article?id=8502 (January 17, 2007).
Newberg, P. R. (2006a), US and Pakistan: An insecure alliance. *Yale Global*, 2/28. Yale Centre for the Study of Globalization. http://yaleglobal.yale.edu/display.article?id=7049 (November 12, 2007)
Nickum, J. E. and K. W. Easter (1994), Metropolitan water: Use conflicts in Asia and the Pacific. Boulder: Westview Press.
Nickum, J. E. and D. Greenstadt (1998), *Transacting a commons: Lake Biwa comprehensive development plan*. In Donahue and Johnston (Eds.), Water, Culture, and Power. Washington, D.C.: Island Press, 1998.
Niemczynowicz, J. (1996), Megacities from a water perspective. *Water International*, 21/4.
Nkrumah, G. (2004, March 3), It must be something in the water: The controversy over Nile water distribution is rearing its ugly head. *Al-Ahram Weekly*.

Norris, P. and R. Inglehart (2004), Sacred and secular: Religion and politics worldwide. Melbourne: Cambridge University Press.

Norton, A. (2004), Leo Strauss and the politics of American empire. New Haven: Yale University Press.

Ohlsson, L. (1995), *Role of water and the origin of conflict*. In L. Ohlsson, (Ed.), Hydropolitics: Conflicts over water as a development constraint. Dhaka: University Press.

Olson, E. (2004, May 18), Unnatural weather, natural disasters: A new UN focus. *NY Times*.

Omvedt, G. (1982), *Class, caste and land in India: An introductory essay*. In G. Omvedt (Ed.), Land, caste and politics in Indian states. Delhi: Author's Guild Publications.

Omvedt, G. (1993), Reinventing revolution: New social movement and the socialist tradition in India. New York: M. E. Sharp.

Omvedt, G. (1994), Dalits and the democratic revolution: Dr. Ambedkar and the Dalit movement in colonial India. New Delhi: Sage Publication.

Onta, I. R. (2001), *Harnessing the Himalayan waters of Nepal: A case for partnership for the Ganges Basin*. In A. K. Biswas and J. I. Uitto (Eds.), Sustainable development of the Ganges-Brahmaputra-Meghna Basin. Tokyo: UN University Press..

Oommen, T. K. (1997), *Social movements in the Third World*. In S. Lindberg and A. Sverrisson (Eds.), Social movement in development: The challenge of globalization and democratization. London: Macmillan Press.

Oommen, T. K. (2005), Crisis and contention in India. New Delhi, Sage Publication.

Organization for Economic Cooperation and Development (1989), Water resource management: Integrated policies. Paris.

Orni, E. and E. Efrat (1973), The geography of Israel (3rd ed.). Jerusalem: The Jewish Publication Society of America.

Osborne, M. (2000), Mekong: Turbulent past, uncertain future; St. Leonard, NSW: Allen and Unwin.

Osborne, M. (2004), River at risk: The Mekong and the water politics of China and South East Asia. Sydney: Lowy Institute.

Osborne, M. (2006), Paramount power: China and the countries of South East Asia. Sydney: Lowy Institute.

Oster, S. (2006, October 25), Beijing hosts big Africa summit as West watches. *Wall Street Journal*.

Oza, N. (1997), Marginalisation, protests and political action: Tribals and SS project. *Economic and Political Weekly*, 32/29.

Pacific Institute (2004), Water scarcity poses threat to global business. Oakland: California. www.pacinst.org.

Pacific Institute (2005), Reinventing globalization. Online update: December 6. Ibid.

Padmanabhan, K. and V. R. Poongavanam (2006), *Blueprint for interlinking rivers*. In S. R. Singh and M. P. Shrivastava (Eds.), River interlinking in India: The dream and reality. New Delhi: Deep and Deep Publications.

Palarisami, K. and R. B. Balasubramaniam (2001), *Reconciling the private well with the community tank.* In Agarwal, A., S. Narayan and I. Khurana (Eds.) (2001), Making water everybody's business: Practice and policy of water harvesting. New Delhi: Centre for Science and Environment.

Panda, M. (2002), *Macroeconomic developments and growth prospects in India.* In K. S. Parikh and R. Radhakrishnan (Eds.), India Development Report 2002. Mumbai: Oxford University Press.

Pandey, G. (1994), The construction of communalism in colonial North India. Oxford: Oxford University Press.

Parikh, K. S. (2002), *Overview: Ten years of reforms, what next?* In K. S. Parikh and R. Radhakrishna (Eds.), India development report 2002. Mumbai: Oxford University Press.

Parker, S. (2005), Water issues becoming increasingly important in US foreign policy. http://www.voanews.com/english/2005-03-17-voa21.cfm?renderfprprint=1 (April 15, 2005).

Patel, A. (1994), *What do the Narmada Valley Tribals want?* In W. F. Fisher. (Ed.), Toward sustainable development? Struggling over India's Narmada River. New York: M. E. Sharp.

Paul, B. K. (1984), Perception of and agricultural adjustment to floods in Jamuna floodplains. *Human Ecology,* 12/1.

Paul, D. H. (1989), Conserving water resources through integrated water management. *Water International,* 14/4.

Paul, S. (1990), Institutional reforms in sector adjustment operations: The World Bank's experience. Discussion Paper No. 92. Washington, D.C.: World Bank.

Pawar, S. N., J. B. Ambedkar and D. Shrikant (Eds.) (2004), NGOs and development: The Indian scenario. Jaipur (India): Rawat Publication.

Pearce, F. (1992), Dammed: Rivers, dams, and the coming world water crisis. London: Bodley Head.

Pearce, F. (2004, April 23), Mekong's vast food resources drying up. *The Canberra Times.*

Pearce, F. (2006), When the rivers run dry: Water, the defining crisis of the twenty-first century. Boston: Beacon Press.

Pemberton, H. (2004), Policy learning and the British government in the 1960s. New York: Palgrave-Macmillan.

Penz, P. (2004), *Development, displacement and international ethics.* In O. P. Mishra (Ed.), Forced migration in the South Asian region. Kolkata: Jadavpur University.

Perrings, C. (2003), Economics of abrupt climate change. *Philosophical Transactions Series A,* 361/1810.

Peters, B. G. (1999), Institutional theory in political science: The new institutionalism. London: Pinter.

Petersen, P. (1999), Grey dawn: The global ageing crisis. *Foreign Affairs,* January–February.

Phadnis, U. (1990), Ethnicity and nation-building in South Asia. New Delhi: Sage.

Phadnis, U. and R. Ganguly (2001), Ethnicity and nation-building in South Asia (Revised edition). New Delhi: Sage.

Plaut, S. (2000), Water policy in Israel. Policy Studies No. 47. Jerusalem: Institute for Advance Strategic and Political Studies.

Pleming, S. (2004), New atlas reveals a dry and unfair world. www.abc.net.au/science/news/enviro/EnviroRepublish_1221019.htm (November 16, 2004).

Postel, S. and C. Flavin (1991), *Reshaping the global economy*. In L. R. Brown, C. Flavin and S. Postel (Eds.), State of the world: A Worldwatch Institute report on progress toward a sustainable society. Sydney: Allen and Unwin.

Postel, S. (1989), Water for agriculture: Facing the limits. Washington, D.C.: Worldwatch Paper No. 93.

Postel, S. (1992), Last oasis (Facing water scarcity). London: W. W. Norton.

Postel, S. (1993), *Water and agriculture*. In P. H. Gleick (Ed.), Water in crisis: A guide to the world's freshwater resources. New York: Oxford University Press.

Postel, S. (1993a), Last oasis: Facing water scarcity. New York: Worldwatch Institute.

Postel, S. (1999), Pillars of sand: Can the irrigation miracle last? New York: W. W. Norton.

Power, J. (2005, February 25), Land reform: The burning issue. *The Statesman*.

Prajapati, R. (1997), Narmada, the judiciary and parliament. *Economic and Political Weekly*, 32/14.

Prasad, N. (2006), *River interlinking in India: A big dream to solve water crisis in distant future*. In S. R. Singh and M. P. Shrivastava (Eds.), River interlinking in India: The dream and reality. New Delhi: Deep and Deep Publications. .

Prasad, R. S. and R. K. Khanna (2002), *Indian experience in development and management of water resources*. In K. Prasad (Ed.), Water resources and sustainable development. New Delhi: Shipra Publications.

Prasad, T., S. K. Bharti and S. Kumar (1987), *Water resources development in India: Its central role in the past and crucial significance for the future*. In W. O. Wunderlich and J. E. Prins (Eds.), Water for the future (Water resources development in perspective). Rotterdam: A. A. Balkema.

Prestowitz, C. (2005), China-India entente shifts global balance. Yale Center for Globalization; *Yale Global on line*, Yale University; http://yaleglobal.yale.edu/display.article?id=5578 (April 16, 2005). Also, see India versus China: The race for the future. *Economic Strategy Institute*, www.econstrat.org/blog/?p=35 (November 19, 2007).

Priscoli, J. D. (2004), What is public participation in water resource management and why is it important? *Water International*, 29/2.

Putnam, R. D. (1988), Diplomacy and domestic politics: The logic of two-level games. *International Organization*, 42/ 3.

Rahaman, M. (2004), Price of free trade: Part 1. http://Yaleglobal.yale.edu/display.article?id=4608 (November 12, 2007).

Rajamani, R. C. (2007, March 12), Cauvery issue sees Deve Gowda in local colour. *The Statesman*.

Rajaraman, I., O. P. Bohra and V. S. Ranganathan (1996), Augmentation of Panchayet resources. *Economic and Political Weekly*, 31/18.

Ram, R. N. (1994), *Benefits of the Sardar Sarovar project: Are the claims reliable?* In W. F. Fisher (Ed.), Toward sustainable development? Struggling over India's Narmada River. New York: M. E. Sharp.

Ramakrishnan, T. (2004, October 8), Sharing water resources. *The Hindu*. http://www.hindu.com/2004/10/08/stories/2004100803561000.htm (October 11, 2004).

Raman, S. (1993, March 6), Bargi oustees plough lonely furrow. *Economic Times*, Ahmedabad.

Ramanathan, U. (1996), Displacement and the Law; *Economic and Political Weekly*, 31/24.

Randel, J. and T. German (Eds.) (1994), Reality of Aid 94: An independent reality of international aid. ICVA/EUROSTEP/ACTIONAID.

Rangley, W. R. (1986), Scientific advances most needed for progress in irrigation. *Philos*; London, Royal Society, A 316.

Rao, R. (1989), Water scarcity hunts world's wettest place. AMBIO, 18/5.

Rao, S. K. (1985), *Rajghat dam: An environmental assessment*. In E. Goldsmith and N. Hildyard, (Eds.), Social and environmental impacts of large dams vol. ii. Camelford, Cornwall, U.K.: Wadebridge Ecological Centre.

Rath, N. (2003), Linking rivers: Some elementary arithmetic. *Economic and Political Weekly*, 38/29.

Rath, N. and A. K. Mitra (1989), Economics of irrigation in water-scarce regions: Study of Maharastra. *Artha Vijnana*, 31/1, Kolkata.

Ravi, R. V. and D. S. Raj (2006), *Empowering rural India: An overview*. In R. V. Ravi, P. S. Ram and D. S. Raj, (Eds.), Empowering rural India: Experiments and experience. New Delhi: Kanishka Publishers.

Ray, B. (1998, January 16 and 17), Farakka Treaty: Can it lead to a better management of water resources in Bangladesh and India? *The Statesman*.

Ray, B. (1999), India: Sustainable development and good governance issues—A case for radical reassessment. New Delhi: Atlantic.

Ray, B. (2004), *Least developing countries of the South Pacific: Future in a globalized economy and unipolar world*. In N. N. Vohra (Ed.), India and Australia—History, culture and society. New Delhi: Shipra Publications.

Ray, B. (2004a), Chassis and wheels, but no engine: Development issues in the South Pacific. *International Journal of Development Issues*, 3/2.

Raymond, R. (1999), A vision for Australia: The Snowy Mountain scheme 1949–1999. Edgecliff: Focus Publishing.

Read, R. (2001), Growth, economic development and structural transition in small vulnerable states. Helsinki: UN University.

Reisner, M. (1986), Cadillac desert: The American west and its disappearing water. London: Secker and Warburg.

Repetto, R. (1986), Skimming the water: Rent-seeking and the performance of public irrigation systems. Washington, D.C.: World Resources Institute.

Revenga, C., J. Brunner, N. Henninger, K. Kassem and R. Payne (2000), Pilot analysis of global ecosystems: Freshwater systems. Washington, D.C.: World Research Institute.

Revkin, A. C. (2007, April 2), Poor nations to bear brunt as world warms. *The New York Times*.

Rich, B. (1994), Mortgaging the earth: The World Bank, environmental impoverishment and the crisis of development. Boston: Beacon Press.

Rich, B. (1994a), Cuckoo in the nest: 50 years of political meddling by the World Bank. *Ecologist*, 24/1.

Rijsbersman, F. (2004, May 15), Water challenge. *The Economist*.

Rizvi, H. A. (1998), *Pakistan: Civil-military relations in a praetorian state*. In R. J. May and V. V. Selochan (Eds.), The military and democracy in Asia and the Pacific. Canberra: ANU E Press.

Rogers, D. J. and S. E. Randolph (2000), Global spread of malaria in a future warmer world. *Science*, 289/5485.

Rogers, P. (1992), Comprehensive water resource management: A concept paper. Washington, D.C.: World Bank.

Rose, N. (2005), Clifford Barclay lecture 2005. London: London School of Economics.

Rosegrant, M. W. (1997), Water resources in the 21st century: Challenges and implications for action. Washington, D.C.: International Food Policy Research Institute.

Rosegrant, M. W. and M. Svendsen (1993), Asian food production in the 1990s: Irrigation investment and management policy. *Food Policy*, 18/1.

Rosegrant, M. W., X. Cai and S. A. Cline (2002), World water and food to 2025: Dealing with scarcity. Washington, D.C.: International Food Policy Research Institute.

Rosenthal, E. and A. C. Revkin (2007, February 2), Panel issues bleak report on climate change. *The New York Times*.

Roy, A. (1999), Cost of living (paperback edition). New York: Modern Library.

Roy, A. (2001), Algebra of infinite justice. New Delhi: Viking.

Roy, S. (2004), Rise to meet the water challenge. http://www.rediff.com/cms/print.jsp?docpath=/news/2004/jul/28waterl.htm (December 11, 2004).

Roy, S. (2007, February 10), Fiscal instability; *Statesman Weekly*.

Rudolph, L. and S. Rudolph (1987), In pursuit of Lakhsmi: The political economy of the Indian state. Chicago: Chicago University Press.

Sah, D.C. (2002), *Some issues relating to voluntary migration*. In S. Tharakan (Ed.), The nowhere people: Responses to internally displaced persons. Bangalore: Books of Change.

Sainath, P. (1998), Everybody loves a good drought: Stories from India's poorest districts. London: Headline Book Publishing.

Saleth, R. M. (1996), Water institutions in India: Economics, law and policy. New Delhi: Commonwealth Publishers.

Saleth, R. M. and A. Dinar (2004), The institutional economics of water: A cross-country analysis of institutions and performances. Cheltenham: Edward Elgar.

Salman, M. A. S. (2003), From Marrakech through the Hague to Kyoto: Has the global debate on water reached a dead end?—Part 1. *Water International*, 28/4.

Salman, M. A. S. (2004), From Marrakech through the Hague to Kyoto: Has the global debate on water reached a dead end?—Part 2. *Water International*, 29/1.

Salman, M. A. S. and L. B. de Chazournes (Eds.) (1998), International watercourses: Enhancing cooperation and managing conflict: Proceedings of a World Bank Seminar. Washington, D.C.: World Bank.

Salman, M. A. S. and K. Uprety (2002), Conflicts and cooperation on South Asia's international rivers. The Hague: Kluwer Law International.

Samad, S. (2004), *Refugees of political crisis in Chittagong Hill Tracts*. In O. P. Mishra (Ed.), Forced migration in the South Asian region. Kolkata: Jadavpur University.

Santarelli, E. and P. Figini (2004), *Some empirical evidence for the developing countries*. In E. Lee and M. Vivarelli (Eds.), Understanding globalization, employment and poverty reduction. New York: Palgrave Macmillan.

Scheffer, D. J. (2003), Beyond occupational law. *American Journal of International Law*, 97/4.

Schindler, D. (2003), *Mountain water: Lifeblood of the prairies*. In B. McDonald and D. Jehl, Whose water is it?—The unquenchable thirst of a water-hungry world. Washington, D.C.: *National Geographic*.

Schwabach, A. (2006), International environmental disputes: A reference handbook. Santa Barbara: ABC-CLIO, Inc.

Schwartz, P. and D. Randall (2003), Abrupt climate change scenario and its implications for US national security. Washington, D.C.: U.S. Dept. of Defense. http://purl.access.gpo.gov/GPO/LPS69716 (November 12, 2007).

Scudder, T. (1990), Victims of development revisited: The political costs of river basin development. *Development Anthropology Network*, 8/1. (New name of the journal is *Development Anthropologist*.)

Sen Sharma, S. (1997), Large dam projects of India? An evaluation. *Science and Culture*, 63.

Sen, A. (1997, November 29), In a speech at the Confederation of Indian Industries [CCII].

Sen, A. (2004), *Tagore and his India*. In Anders Hallengren (Ed.), Nobel laureates— In search of identity and integrity—Voices of different cultures. Singapore: World Scientific Publishing.

Sengupta, N. (1985), Irrigation: Traditional vs. modern. *Economic and Political Weekly*, 20/45, 46, and 47.

Serageldin, I. (1995), Towards sustainable management of water resources. Washington, D.C.: World Bank.

Seshan, T. N. (1995), Degradation of India. New Delhi: Viking.

Shah, T. (1993), Groundwater markets and irrigation development. Bombay: Oxford University Press.

Shankar, R. (2004), *Voluntary associations in social change initiators*. In S. N. Pawar, J. B. Ambedkar and D. Shrikant (Eds.), NGOs and development: The Indian scenario. Jaipur: Rawat Publication.

Shaplen, J. T. and J. Laney (2004, July 12), China trades its way to power. *The New York Times*.

Sharan, R. (1997), Surangi irrigation project: Oustees left in lurch. *Economic and Political Weekly*, 32/9 and 10.

Sharma, S. D. (2003), *Indian politics*. In S. Ganguly and N. DeVotta (Eds.), Understanding contemporary India. Boulder: Lynne Rienner Publishers.

Sharma, S. D. (2004), Budgeting interlinking of rivers. Delhi: The Ecological Foundation. The updated version is published in the Foundation website, http://countercurrents.org/ensharma.htm.

Shastri, S. (2001), *Institutional arrangements in federal systems and their role in resolving intra-state conflicts*. In S. J. George (Ed.), Intra and inter-state conflicts in South Asia. New Delhi: South Asian Publishers.

Sherpa, D. M. (1994), Living in the middle: Sherpas of the middle-range Himalayas. Prospect Heights: Waveland Press.

Shih, Chih-yu (1990), Spirit of Chinese foreign policy: A psychological view. Houndmills, Hampshire: Macmillan.

Shiklomanov, I. A. and N. V. Penkova (2003), *Methods for assessing and forecasting global water use and water availability*. In I. A. Shiklomanov and J. C. Rodda (Eds.), World water resources at the beginning of the 21st century. Cambridge, U.K.: Cambridge University Press.

Shiklomanov, I. A. (1990), Global water resources. *Nature and Resources*, 26/3.

Shiklomanov, I. A. (1993), *World freshwater resources*. In P. H. Gleick (Ed.), Water in crisis: A guide to the world's freshwater resources. New York: Oxford University Press.

Shiva, V. (2003), Sujalam: Living waters—The impact of the river linking project. New Delhi: Navdanya.

Shonefield, A. (1965), Modern capitalism. London: Oxford University Press.

Short, R. and C. McConnell (2001), Resource management technical report 202: Agriculture department. Perth: Government of Western Australia.

Shrivastava, M. P. (2006), *Socio-economic dreams of river interlinking in India*. In S. R. Singh and M. P. Shrivastava (Eds.) (2006), River interlinking in India: The dream and reality. New Delhi: Deep and Deep Publications.

Singh, G. (2003), Global corruption report. Berlin, Transparency International. http://unpan1.un.org/intradoc/groups/public/documents/APCITY/UNPAN008446.pdf (November 16, 2007).

Singh, K. B. (1979), *Water Management at the farm level in India*. In Ghassemi, F., A. J. Jakeman and H. A. Nix (Eds.) (1995), Salinisation of land and water resources: Human causes, extent, management and case studies. Sydney: University of NSW Press.

Singh, R. (2003), Interlinking of rivers. *Economic and Political Weekly*, 38/40.

Singh, S. (1997), Taming the waters: The political economy of large dams. New Delhi: Oxford University Press.
Singh, S. R. and M. P. Shrivastava (Eds.) (2006), River interlinking in India: The dream and reality. New Delhi: Deep and Deep Publications.
Singh, S. R. (2006), *Interlinking of rivers and mission 2020.* In S. R. Singh and M. P. Shrivastava (Eds.), River interlinking in India: The dream and reality. New Delhi, Deep and Deep Publications.
Singh, S., A. Kothari and K. Amin (1992), *Evaluating major irrigation projects in India.* In E. G. Thukral (Ed.), Big dams, displaced people: Rivers of sorrow, rivers of change. New Delhi: Sage.
Singh, Santosh (2004, July 13), Where was Medha when Harsud was sinking? *The Statesman.*
Sinha, B. K. (1996), Draft national policy for rehabilitation: Objectives and principles. *Economic and Political Weekly,* 31/24.
Smith, W. (1989), Sustainable use for water in the 21st century. *AMBIO,* 18/5.
Sneeuwjagt, R. (2005: March 16), Controlled burning opponents fire up. http://www.abc.net.au/southwestwa/stories/s1334649.htm (November 20, 2007).
Sobhan, R. (2000), *Future of SAARC: Towards building a South Asian community.* In D. K. Dutta (Ed.), Economic liberalization and institutional reform in South Asia: Recent experiences and future prospects. New Delhi: Atlantis Publishers.
Social Science Council of Canada (1995), Water 2020, Report 40. Ottawa.
Soltan, K. E. (1996), *Institution building and human nature.* In K. E. Soltan and S. L. Elkin (Eds.), Constitution of good societies. University Park: Pennsylvania University Press.
South Australian Department for the Environment (1980), Generic guidelines for an environmental impact statement. Adelaide.
South Australian Department of Housing and Urban Development, Environment Australia (1997), Guidelines for an environmental impact statement on the proposed expansion of the Olympic dam. Adelaide.
South Pacific Applied Geoscience Commission (1998), Desalinisation: A technical appraisal for its application in Pacific Island countries.
Sowani, V. (2006), *Interlinking of rivers: Significance and prospects.* In S. R. Singh and M. P. Shrivastava (Eds.), River interlinking in India: The dream and reality. New Delhi: Deep and Deep Publications.
State Salinity Council (2000), The salinity strategy: Natural resource management in Australia. Perth: Government of Western Australia.
Stephens, C. and S. Bullock (2004), *Civil society and environmental justice.* In P. Gready (Ed.), Fighting for human rights. London: Routledge.
Stern, N. H. (2006), Stern review: Economics of climate change (Executive summary). London: H. M. Treasury.
Stiglitz, J. (2007, January 10), Stiglitz pat in times of turbulence. *The Statesman.*
Strauss, S.D. (2002), Complete idiot's guide to world conflicts. ALPHA—A Pearson Education Company.

Strong, M. and G. Goransson (1991), Foreword. Water: The international crisis. Cambridge, Mass.: MIT Press.

Struck, D. (2006, May 5), Climate change drives disease to new territory. *The Washington Post.*

Svensen, H., S. Planke, A. Malthe-Sorenssen, B. Jamtveit, R. Myklebust, T. Rasmussen Eidem, et al. (2004), Release of methane from a volcanic basin as a mechanism for initial Eocene global warming. *Nature*: 429/6991.

Swaminathan, M. S. (1986), Foreword. In proceedings of the conference on common property resource management. Washington, D.C.: National Academies Press.

Swaminathan, M. S. (2004, July 18), End greed revolution. *The Statesman.*

Swan, B. (1987), Sri Lankan mosaic: Environment, man, continuity and change. Colombo: Marga Institute.

Tanzi, A. and M. Arcari (2001), UN convention on the law of international watercourses. The Hague: Kluwer Law International.

Tata Services Ltd. (2002–3), Statistical Outline of India. Mumbai.

Thakur, T. (2006, April 29), Green revolution's wounded warriors. *Statesman Weekly.*

Tharakan, S. (Ed.) (2002), The nowhere people: responses to internally displaced persons. Bangalore: Books of Change.

Thompson, M. (1995), *Disputed facts. A countervailing view from the hills.* In G. P. Chapman and M. Thompson Chapman (Eds.) (1995), Water and the quest for sustainable development in the Ganges valley. New York: Mansell Publishing.

Thorner, D. and A. Thorner (1962), Land and labour in India. Bombay: Asia Publishing House.

Tiwary, M. (2004), Participatory forest policies in India. Ashgate: Aldershot.

Tortajada, C. (2003), Workshop on Integrated Water Resource Management for South and South East Asia. *Water International*, 28/4.

Transparency International (2003), Annual Report 2003:160.

Trevin, J. O. and J. C. Day (1990), Risk perception in international river basin management: The Plata Basin example. *Natural Resource Journal*, (winter issue).

Truell, P. and L. Gurwin (1992), False profits: The inside story of BCCI, the world's most corrupt financial empire. Boston: Houghton Miffin.

UN (1986), Ground water in continental Asia: Natural resources/water series No.15. New York: U.N.

UN (1987), Our common future, WCED. Oxford: Oxford University Press.

UN (1989), E/ECE 1197, ECE/ ENVWA/12: Charter on ground-water management. Economic Commission for Europe.

UN (1989a), Ground water development and management in developing countries: 25 years of UN activities 1963–88. New York: UN.

UN (1995), Guidebook to water resources, use and management in Asia and the Pacific, volume one, New York.

UN (1997), Convention on the law of the non-navigational uses of international watercourses. UNGA A/51/869. New York: UN.

UN (1998), Guiding principles on internal displacements: Submitted by the UN Secretary General on internally displaced persons to the Commission on Human Rights at the 54th Session—E/CN.4/1998/53/Add.2.

Uliveppa, H. H. (2006), *Interlinking of rivers in India: Problems and prospects*. In S. R. Singh and M. P. Shrivastava (Eds.), River interlinking in India: The dream and reality. New Delhi: Deep and Deep Publications.

Ullman, R. H. (1983), Redefining security. *International Security*, 8/1.

UNAIDS (2006), Report on the global AIDS epidemic: Executive summary. A UNAIDS 10th anniversary special edition. Geneva: UNAIDS Information Centre, WHO.

UNCHS (1991), Water and sustainable urban development and drinking water supply and sanitation in the urban government; Background paper for the Dublin water and the environment Conference. New York: U.N.

UNCSD (1997), Comprehensive assessment of the freshwater resources of the world: Report of the Secretary General. Geneva: UN.

UNCTAD (1989), Africa's commodity problems: Towards a solution. Task force on UN-PAAERD, Geneva: UN.

UNDP (1990, 2001, 2002, 2003, 2004), Human Development Reports. Oxford University Press.

UNEcoSoC (2004), 12th Session April 14–30, 2004; E/CN.17/2004/4.

UNEP (2001), Rainwater harvesting and utilisation: UNEP-DTIE-IETC/Sumida City Government / People for promoting rainwater utilisation.

UNWWDR (2003), Water for people, water for life. New York: UNESCO Publications: Berghan Books.

Upadhyaya, S. P., and B. N. Sapkota (1985), *Water resources development in Nepal*. In Ali, Md., G. E. Radosevich and A. A. Khan (Eds.) (1987), Water resources policy for Asia. Rotterdam: A. A. Balkema.

Upreti, B. C. (1993), Politics of Himalayan river waters: An analysis of the river water issues of Nepal, India and Bangladesh. Jaipur: Nirala Publications.

US Water News (1995, January issue), Deforestation creates drought in wettest spot of the world.

Usher, A. D. (Ed.) (1997), Dams as aid: A political anatomy of Nordic development thinking. London: Routlege.

Utton, A. E. (1987), *Selected issues of groundwater management*. In Ali, Md., G. E. Radosevich and A. A. Khan (Eds.) (1987), Water resources policy for Asia. Rotterdam: A. A. Balkema.

van der Gaag, J. (2002), *Standards of care: Investments to improve children's educational outcomes in Latin America*. In M. E. Young (Ed.), From early child development to human development: Investing in our children's future. Washington, D.C.: World Bank.

Varghese, B. G. (1996), *Towards an Eastern Himalayan rivers concord*. In A. K. Biswas and J. I. Uitto (Eds.), Sustainable development of the Ganges-Brahmaputra-Meghna basin. Tokyo: UN University Press.

Varma, S. P. (1986), *Crisis in South Asia: An overview*. In U. Phadnis, S. D Muni and K. Bahadur (Eds.), Domestic conflicts in South Asia. New Delhi: South Asian Publishers.

Vellinga, P. and S. P. Leatherman (1989), Sea level rise: Consequences and policies. *Climatic Change*, 15/ 1–2.

Vidal, J. (1995, August 8), The water bomb. *The Guardian*.

Vincent, M. (2001), Response strategies: The need to involve the displaced. Norwegian Refugee Council Report of a seminar held in 2001: Response strategies of the internally displaced: Changing the humanitarian lens. Oslo.

Vira, B. and S. Vira (2004), *India's urban environment: Current knowledge and future possibilities*. In T. Dyson, R. Cassen and L. Visaria (Eds.) (2004), Twenty-first century India: Population, economy, human development and the environment. New Delhi: Oxford University Press.

Vira, B., R. Iyer and R. Cassen (2004), Water. In T. Dyson, R. Cassen and L. Visaria. Ibid.

Vivarelli, M. (2004), *Globalization, skills and within-country income inequality in developing countries*. In E. Lee and M. Vivarelli (Eds.) (2004), Understanding globalization, employment and poverty reduction. New York: Palgrave Macmillan.

Volcker, P. (2007, February 19), India responded to findings. *The Statesman*.

Vorosmarty, C. J., P. Green, J. Salisbury and R. B. Lammers (2000), Global water resources: Vulnerability from climate change and population growth. *Science*, 289/5477.

Wade, R. (1985), Market for public office: Why the Indian state is not better at development. *World Development*, 13/4.

Wade, R. (1988), Village republics: Economic conditions for collective action in South Asia. Cambridge: Cambridge University Press.

Waller, D. H. (1989), Rain water—An alternative source in developing and developed countries. *Water International*, 14/1.

Wapner, P. (2000), *Transnational politics of environmental NGOs: Governmental, economic and social activism*. In P. S. Chasek (Ed.) (2000), The global environment in the twenty-first century: Prospects for international cooperation. Tokyo: UN University Press.

Ward, C. (1997), Reflected in water: A crisis of social responsibility. London: Cassell.

Waterbury, J. (1997), Between unilateralism and comprehensive accord: Modest steps toward cooperation in international river basins. *International Journal of Water Resources Development*, 13/3.

Weiner, M. (1987), *Political change: Asia, Africa and the Middle-East*. In M. Weiner and P. Huntington (Eds.), Understanding political development: An analytical study. Boston: Harper Collins.

Weiner, M. (1987a), *India: Postcolonial democracies*. In M. Weiner and E. Ozbudun (Eds.), Competitive election in developing countries. Durham: Duke University Press.

Weiner, M. (1989), Indian paradox: Essays in Indian politics. New Delhi: Sage.

Weiner, M. and R. Kothari (Eds.) (1965), Indian voting behaviour: Studies of the 1962 general election. Calcutta: Firma K. L. Mukhopadhayay.

Wescoat, J. L., Jr., and G. F. White (2003), Water for life: Water management and environment policy. New York: Cambridge University Press.

West Bengal Academy of Science and Technology (2003), Proceedings of the workshop on interlinking of rivers. Calcutta: Indian Institute of Chemical Biology.

Whipple, W., Jr. (1994), New perspectives in water supply. London: Lewis Publishers.

Whitcombe, E. (1972), Agrarian conditions in Northern India, vol. 1: The United Province under British Rule 1860–1900. Barkley: University of California Press.

White, C. (2004), Three Gorges Dam. *Discovery on line*. www.ccds.charlotte.nc.us/History/China/02/Cwhite/Cwhite.htm (November 12, 2007).

Wigmore, L. (1968), Struggle for the Snowy: The background of the Snowy Mountain scheme. Melbourne: Oxford University Press.

Wilks, A. and Hildyard, N. (1994), Evicted! World Bank and forced resettlement. *Ecologist*, 24/6.

Williams, P. B. (1989), Adapting water resource management to global climate change. *Climatic Change*, 15/1–2 (special issue).

Williams, P. B. (1993), *Sedimentation analyses*. In M. Barber and G. Ryder (Eds.), Damming the Three Gorges (2nd edition). London: Probe International and Earthscan.

Winpenny, J. (1994), Managing water as an economic resource. London: Overseas Development Institute and Routledge.

Wolf, A., J. A. Natharius, J. J. Danielson, B. S. Ward and J. K. Pender (1999), International river basins of the world. *International Journal of Water Resources Development* 15/4.

Wolff, G. and E. Hellstein (2005), Beyond privatization: Restructuring water systems to improve performance. California: Pacific Institute. http://www.pacinst.org/reports/beyond_privatization/BeyondPrivate_ExSum.pdf.

Wood, R. E. (1986), From Marshall Plan to debt crisis: Foreign aid and development choices in the world economy. Berkeley: University of California.

World Bank (1991), India irrigation sector review. Washington, D.C: World Bank.

World Bank (1992), World development report: Development and the environment.

World Bank (1997), India—Achievements and challenges in reducing poverty.

World Bank (1997a), World development report: The state in a changing world.

World Bank (2000 and 2002), World development indicators.

World Health Organization (2003), Guidelines for safe recreational water environments, volume 1: Coastal and fresh waters. Geneva.

World Meteorological Organization (1991), Climate change: The IPCC response strategies. Washington, D.C.: Island Press.

World Meteorological Organization (1997), Comprehensive assessment of the freshwater resources of the world. Geneva: WHO.

World Resources (1992), World resources 1992–93: A report in collaboration with UNEP and UNDP. New York: Oxford University Press.

World Resources (1998), World resources 1998–99: A guide to the global environment, environmental change and human health. New York: Oxford University Press.

Worster, D. (1986), Hoover dam: A study in domination. In E. Goldsmith and N. Hildyard (Eds.), Social and environmental impacts of large dams, vol. ii. Camelford, Cornwall, U.K.: Wadebridge Ecological Centre.

Yahuda, M. (2002), *China and regional cooperation*. In K. E. Brodsgaard and B. Heurlin (Eds.), China's place in global geopolitics: International, regional and domestic challenges. New York: RoutledgeCurzon.

Yasin, M. and P. K. Sengupta (2004), Indian politics: Processes, issues and trends. New Delhi: Kanishka Publishers and Distributors.

Yaswant, S. (1993, July 30), Bijasen and beyond: Driven away by dams. *Frontline*.

Zaman, M. Q. (1982), Crisis in Chittagong Hill Tracts—Ethnicity and integration. *Economic and Political Weekly*, 17/3.

Zedillo, E. (2007), Debating the price of global warming. http://yaleglobal.yale.edu/display.article?id=8930 (March 22, 2007).

Zhaoxin, W. (1990), *Reciprocal water transfer between groundwater and surface water conditions of water resources development*. In U. Shamir and C. Jiaqui (Eds.), Hydrological basis for water resources management. *Wallingford, International Association of Hydrological Sciences Publication* No. 197.

Zimmerman, J. D. (1966), Irrigation. New York: John Wiley & Sons.

Further Reading

Adams, J. (1996), Cost-benefit analysis: The problem, not the solution. *Ecologist*, 26/1.
Agarwal, A. (Ed.) (1997), Challenge of the balance. New Delhi: Centre for Science and Environment.
Ahmed, A. S. (1997), Jinnah, Pakistan and Islamic identity: The search for Saladin. London: Routledge.
Alexandratos, N. (Ed.) (1995), World agriculture: Towards 2010. Rome: FAO.
Ali, C. R. (1949), Muslim minority in India and the Dinian mission to the UNO. Lahore: All-Dinia Milli liberation movement.
Allan, Tony (1998), War and water. Geneva: ICRC Forum.
Amery, H. A. and A. T. Wolf (2000), Water in the Middle East: A geography of peace. Austin: University of Texas Press.
Anderson, C. W. (1996), *How to make a good society*. In K. E. Soltan and S. L. Elkin (Eds.), The contribution of good societies. Pennsylvania: Pennsylvania State University Press.
Banks, S. and T. C. Muller (Eds.) (1999), Political handbook of the world. Binghamton, N.Y.: CSA Publications.
Benedini, M. (1989), Optimization in combined use of groundwater and surface water resources. In W. James and J. Niemczynowicz (Eds.), Water, development and the environment—Proceedings of the international symposium on water, development and the environment. Boca Raton: Lewis Publishers. http://www.loc.gov/catdir/enhancements/fy0744/92005179-d.html.
Biltonen, E. and J. A. Dalton (2003), A water-poverty accounting framework: Analysing the water-poverty link. *Water International*, 28/4.

Bose, S. (1994), States, nations, sovereignty: Sri Lanka, India and Tamil Elam movement. New Delhi, Sage.
Branch, M. C. (1998), Comprehensive planning for the 21st century: General theory and principles. Westport: Praeger.
Brown, M. E. and R. N. Rosecrance (1999), Case for conflict prevention. In M. E. Brown and R. N. Rosecrance (Eds.), Costs of conflict prevention and cure in the global arena. Lanham, Md.: Rowman & Littlefield Publishers.
Buchanan, J. (1990), Domain of constitutional economics. *Constitutional Political Economy*, 1/1.
Buzan, B. (2004), *Conclusions: How and to whom does China matter?* In B. Buzan and R. Foot (Eds.), Does China matter? A reassessment. London: Routledge.
Calhoun, C. (Ed.) (1992), Habermas and the public sphere. Cambridge, Mass.: MIT Press.
D'Monte, D. (1985), Temples or tombs. New Delhi: Centre for Science and Environment.
de Soysa, I. (2001), Paradise is a bazaar? Greed, creed, grievance and governance. Helsinki: UN University,
Diamond, J. (2005), Collapse: How societies choose to fail or succeed. New York: Viking.
Dorgi, J. and A. K. Pradhan (1987), *Water resources development in Bhutan*. In Ali, Md., G. E. Radosevich and A. A. Khan (Eds.), Water resources policy for Asia. Rotterdam: A. A. Balkema.
Dreze, J., M. Samson and S. Singh (1997), Dam and the nation. New Delhi, Oxford University Press.
Ekeh, P. (1975), Colonialism and the two publics in Africa: A theoretical statement. *Society and History*, 17.
Fair, C. (2007, March 16), Bangladesh's fragile democracy (Interview with H. Vatsikopoulos of the *Australian Broadcasting Commission's Australia network—Asia Pacific focus program*. http://australianetwork.com/focus/s1874872.htm (November 12, 2007).
Ferguson, I., M. Lavalette and E. Whitmore (Eds.) (2005), Globalisation, global justice and social work. London: Routledge.
Flyvberg, B. and V. C. Petersen (1983), Planning theory: Theoretical considerations on the analysis of public policy and planning. Aalborg: University Press. Public Planning Series No. 15.
Friedrich, C. J. (1972), Pathology of politics: Violence, betrayal, corruption, secrecy and propaganda. New York: Harper and Row.
Galtung, J. (1985), Development theory: Notes for an alternative approach. Berlin: International Institute for Environment and Society.
Gellner, E. (1999), *Coming of nationalism, and its interpretation: The myth of nation and class*. In S. Bowles, M. Franzini and U. Pagano (Eds.), The politics and economics of power. London: Routledge.

Gleick, P. H. (1989), Climate change and international politics: Problems facing developing countries. *AMBIO*, 18/6.

Hatsuko, H. (2004, July 30), Struggle over the Arase dam: Japan's first dam removal begins. *Shukan Kinyobi*. Translated from Japanese by Lebowitz.

Homer-Dixon, T. F. and J. H. Boutwell (1993), Environmental change and violent conflict. *Scientific American*, 268/2.

Hudson, J. and S. Lowe (2004), Understanding the policy process: Analysing welfare policy and practice. Bristol: The Policy Press.

Huntington, S. P (1996), The clash of civilizations and the remaking of world order. New York: Simon Schuster.

IIMI (1992), Developing environmentally sound and lasting improvements in irrigation management: The role of the institute research. Colombo.

Jalal, A. (1985), Sole spokesman: Jinnah, the Muslim League and the demand for Pakistan. New York: Cambridge University Press.

Jetly, N. (1986), *India and the domestic turmoil in South Asia*. In U. Phadnis, S. D. Muni and K. Bahadur (Eds.) (1986), Domestic conflicts in South Asia. New Delhi: South Asian Publishers.

Just, R. E. and S. Netanyahu (1998), Conflict and cooperation on transboundary water resources. Boston: Kluwer Academic Publishers.

Lama, M. P. (1998), *Issues in harnessing economic resources in South Asia*. In B. D. Ghoshal (Ed.), ASEAN and South Asia development experience. New Delhi: Sterling.

Major, D. C. and H. E. Schwarz (1990), Large-scale regional water resources planning. Boston: Kluwer Academic Publishers.

Mandel, R. (2004), Security, strategy, and the quest for bloodless war. Boulder: Lynne Rienner.

McKinley, M. (1989), *India and the sub-continent*. In G. Klinworth (Ed.), China's crisis: The international implications. Canberra: Australian National University.

Nye, J. S. Jr., and J. D. Donahue (Eds.) (2000), Governance in a globalizing world. Washington D.C.: Brooking Institute Press.

Osborne, M. (2001), Mekong. Sydney: Allen and Unwin.

Ostrom, E. (1996), *Covenants, collective action and common-pool resources*. In K. E. Soltan and S. L. Elkin (Eds.), Constitution of good societies. University Park: Pennsylvania University Press.

Panth, N. (1979), Some aspects of irrigation administration (a case study of the Kosi project). Patna: Anugraha Narayan Sinha Institute of Social Studies.

Prokopy, L. S. (2005), Relationship between participation and project outcomes: Evidence from rural water supply projects in India. *World Development*, 33/11.

Roy, A. (2000), Cost of living. *Frontline*, 17/03.

Sands, P. (2005), Lawless world. New York: Viking.

Schiavo-Campo, S. (Ed.) (1999), Governance, corruption and public financial management. Manila: Asian Development Bank.

Scruton, R. (2000), Who, what and why? Discussion paper 113. London, Institute of Economic Affairs.

Shah, T. (1985), Transforming ground water markets into powerful instruments of small farmer development: Lessons from the Punjab, Uttar Pradesh and Gujrat. *Irrigation Management Network Paper 11d*. London: ODI.

Shah, T. (1989), Efficiency and equity impacts of groundwater markets: A review of issues, evidence and policies. Anand: Institute of Rural Management. Research Paper No. 8.

Sheehan, P. (2003), Electronic whorehouse. Sydney: Macmillan by Pan Macmillan.

Shiklomanov, I. A. (2004), World water resources and their use: A joint study by the Russian state hydrological institute. St. Petersburg: UNESCO.

Singh, P. and A. Bhandarkar (1997), IAS profile: Myths and realities. New Delhi: Wiley Eastern.

Stone, I. (1984), Canal irrigation in British India: Perspectives on technological change in a peasant economy. Cambridge: Cambridge University Press.

Taimni, B. K. (2004), War on poverty: For taking poor to portals of civil life. New Delhi: A. P. H. Publishing Corporation.

Vaidyanathan, A. (1997), *Tanks and tank irrigation*. In Agarwal, A. (Ed.) (1997), Challenge of the balance. New Delhi: Centre for Science and Environment.

van de Laar, A. (1980), World Bank and the poor. Boston: Martinus Publishing.

Waldrop, M. M. (1992), Complexity: The emerging science at the edge of order and chaos. New York: Simon and Schuster.

World Commission on Dams (2000), Dams and development: A new framework for decision making. London: Earthscan.

Index

Act of the Congress of Vienna, 51
Agenda 21, 6, 19, 19n3
agriculture: freshwater use, 3, 33–34, 59, 80, 133, 180; practices, 36, 135, 153; productivity, 35, 84, 86, 87n22, 107, 132; sector performance, 120–21. *See also* irrigation
aid, 9, 10, 20n, 12, 25, 74, 112, 123, 158, 186
Andhra Pradesh, 5, 44–45, 68, 132, 154
arable land, 79, 80, 81
Arctic Climate Impact Assessment Group, 110
arsenic, 69
Asian Development Bank, 10, 72
Athens Resolution, 53
Australia, 54, 99, 111, 132, 152, 159

Babhli irrigation project, 132
Baglihar project, 59
Bangladesh, xxv, 18, 53, 55–57, 60, 65, 69, 71, 79, 80, 83, 92, 95, 96–98, 110, 117, 141–45, 151, 153
Barcelona Convention, 53–54
Bargi dam, 72

Bay of Bengal, 56, 92, 99, 101, 179
Belgrade Resolution, 53
Bhutan, xvii, xviii, xxv, 5, 10, 15, 28, 55–56, 58–59, 65, 92, 97, 143–45, 151, 155, 175
Bihar, 5, 29, 44, 95, 97
bilateralism, 28, 55, 60, 144, 150–51, 157, 178, 186
Brahmaputra, xvii, xxv, 13, 15–16, 18–21, 29, 57, 73, 92–93, 98, 111, 142–43, 150, 156, 182
Britain, 54, 83, 132, 134, 158, 165, 177
Bhutan, xvii–xviii, xxv, 5, 10, 15, 28, 55–56, 58–59, 65, 92, 97, 143–45, 151, 154, 175
Bundelkhand, xiii–xv

California, 36, 107
Canada, 5, 8, 50, 94, 106–7
Cauvery, 27, 91, 130, 139n3
central government, 28, 44, 91, 119, 128–29, 134, 136, 138, 170
Central Water Commission, 37
Cherapunji, 4, 73
child mortality rates, 117

China: economy, 65, 116, 118, 122, 156; environment, 67, 106, 111; international relations, 59, 106, 122, 143–45, 150, 155–56, 159, 180–83; resource use, xxviin9, 65, 82, 116, 179–80; rivers, 92, 142–43, 156–57, 179–80. *See also* water sharing
climate change, xvii, 7, 16, 24, 28, 30–31, 34, 50, 67, 72, 79, 93–94, 96–97, 105–12, 113n3, 123, 141, 146, 149–50, 162, 183
coalition government, 120, 130, 132, 134, 153
coastal erosion, 99
colonialism, xvii–xviii, 10, 51–52, 54, 60, 61n1, 84, 158–59, 165
compensation, xiii–xiv, 39, 55
congress, 130, 133–34, 139n7
Constitution, 28, 39, 41–42, 127–29, 132, 134, 136–38
contamination, 8, 36, 69–70, 75–76, 143
Convention on the Rhine, 50
coral reefs, 111
corruption, 18, 34, 38–39, 46n5, 46n7, 122, 124n3, 161, 165, 170, 176–77
cost benefit analysis, xiv, 8, 68, 124
Council of Parties, 106
Council of the League of Nations, 54
customary law, 44, 49

Dalits, 40, 43
dams: benefits, xxiv, 8, 9, 74, 149; conflicts, xxv, 16, 21, 57–58, 71–74, 156, 165–69, 180, 183; construction, xxiv, 8, 29, 32, 37, 67, 70–75, 90, 93–94, 107–8, 111–12, 116, 142, 145, 149, 151, 183; environmental impacts, 73, xii–xiii, 29, 36, 83–84, 98–99, 107–8, 110, 153, 180; health effects, 36, 83–84; social impacts, 94, 98, 149, 168. *See also* population; environmental impact assessment

Damodar Valley project, 8
Danube River, 28, 50
Daudhan dam, xii–xiii
dead zones, 99
Deccan Plateau, 69, 75, 92
deficit, 116, 118–19, 124, 129
deforestation, 6, 73–74, 80, 98, 143
demand. *See* freshwater
dengue, 35
desalinization, 67, 77, 85
development agencies. *See* aid
diseases. *See* health
displacement of people. *See* population
donor policies. *See* aid
drinking water, xii, xiv, 6, 19n2, 25, 63, 68–70, 79, 99, 128

economic costs, 80, 90, 95, 115–16, 131, 146
economy, xxi, 29, 32, 35, 59, 76, 79, 85, 107–08, 117, 121–23; global, 123, 156
education, 25, 30, 34, 78, 84, 116–18, 158, 161
Egypt, 60
employment, xiv, 42, 121–22, 152, 173, 178
energy, 77, 79, 85, 107–8, 116, 180
environmental impact assessment, 41–42, 55, 75, 90, 95–97, 105–6, 108–12, 123, 153, 162, 171
environmental impacts, 8, 9, 13, 27–29, 32, 34, 36–37, 42, 50–51, 72, 76, 79, 83–85, 92–93, 99–100, 109–11, 132–33, 152. *See also* deforestation; climate change
erosion, 73–74, 98, 107, 143
ethnic dimensions, 135, 157–58, 167, 170, 180
European countries, 50–51, 95, 107, 109, 178, 180
evaporation, 4, 5, 16, 66, 68, 83–84, 132, 150, 184

Farakka, 18–19, 21, 27, 56–57, 83, 98, 142, 145, 151, 177–78
flooding, 70–71, 141–43, 145
Food and Agricultural Organisation, 26, 64
France, 8, 51, 54
freshwater: availability, xi–xii, xvii, xxi, 3, 4, 7, 15, 17, 24, 32, 45, 56, 66–67, 89; demand, 5, 10, *11*, *12*, 31, 52, 63–64, 65, 66–67, 99, 155; distribution, 3, 89, 91–92, 127–28, 141–44; policy, xviii, xxiii–xxv, 17–18, 30, 42, 67, 141–42, 149, 171. *See also* sustainable use of water; water policy; water sharing
Friendship Treaty, 59, 145
funding, 9, 41–42, 74, 100, 109, 120–23

Gandaki treaty, 56, 58
Ganges, xxv, 18, 56–57, 83, 92–93, 98–99, 110–11, 133, 141–43, 150, 156
Gangetic, 95, 97–99, 101n14, 142
Gangotri glacier, 28, 107, 110
Geneva Convention, 53
glaciers, 16, 20n17, 79, 93, 107, 110–11, 141, 143, 150. *See also* Gangotri glacier
governance, xxii, 18, 105–6, 122, 124, 127–39, 145, 149, 158, 175, 183
government economic policies, xxii–xxiv
grassroots contributions, 31, 34, 87n16, 124, 127, 152, 158–59
Green Revolution, xi
greenhouse effect. *See* climate change
gross domestic product, xxii, 115–16, 119, 123, 146, 167
groundwater: contamination, 34, 36, 51, 55, 67, 69, 75–76; harvesting, 26, 34, 75–76, 79, 85; trans-boundary, 53, 55; value of, 55, 69. *See also* freshwater

Haiti, 110
Harmone Doctrine, 52
Harsud dam, 39
health, 27, 32, 35–37, 61, 68–69, 78, 116–18, 132, 150, 155, 176
Helsinki Convention. *See* rivers
Himalayan, xxi, xxiv, *xxxiv*, 5, 13, 28–29, 74, 79, 89–90, 92–93, 107–8, 115, 141, 150, 183
Hirakud dam, 70
HIV, 116–17
holy water places, 23
Human Development Index, 118
human rights, 30, 44, 123, 162, 163n1, 175, 180
Human Rights Commission, 44, 118
hydroelectricity, 8, 59, 124, 165
hydrological projections, 30

Independence, 24, 37, 57, 60, 70–71, 84, 89, 130, 133–34, 137, 149, 158, 169
Indian Forest Act, 43
Indian Public Accounts Committee, 71
Indira Gandhi canal system, 37, 40
Indonesia, 54, 116, 122
Indus river system, xxv, 16, 56, 74, 142, 184
Indus treaty, 59
Institute of International Law, 52
institutional reform, 57, 95, 97, 127, 134–39
integrated water management, xxiv, 45n3, 57, 132, 137, 150, 154–55, 158, 171
Intergovernmental Panel on Climate Change, 94, 105–6, 108–9
International Commission on Large Dams, 74
International Court of Justice, 49, 157
International Law Association, 52
International Law Commission, 52
International Monetary Fund, 5, 20n9

228 ～ Index

interstate water disputes, 129, 130–35
Interstate Water Disputes Act, 129, 133, 136
investment, 59, 67, 78, 80, 83–84, 86, 90, 109, 112, 117–18, 120–22
irrigation: canal, xiii, 83–84, 132–33, 158; and freshwater demand, 10, 16, 18, 26, 29, 34, 66, 79, 131–32, 135, 141, 161; pressures, xii, xiv, 25, 32, 40, 59, 80, 89, 109, 121, 128; productivity, xiv, 9, 20n11, 35, 71–72, 83–84, 85, 94, 116, 120, 131
Italy, 8, 54

Japan, 8, 37, 54, 106, 112, 123, 156, 159,
judicial activism, 137–38

Ken-Betwa link, xii, xiv, 90
Ken Ghariyal Sanctuary, xii
Kenya, 60
Kolkata, 18, 57, 79, 99
Kosi, 56, 58, 144
Kyoto Protocol, 50, 77, 106, 109

Land Acquisition Act, 43–44
Land Acquisition Amendment Bill, 43
Land Acquisition, Rehabilitation and Resettlement Bill 2000, 42
land reform, 34–35, 94, 135, 159
local government, 119, 127–28

Madhesi, 58
Madrid Declaration, 53
Magna Carta, 61
Mahakali Treaty, 56, 58
malaria, xiii, 35
Malaysia, 54
mangroves, 98–99, 102n17
Mar del Plata, 6, 63
marginalized communities, 9, 13, 29, 38, 40–41, 44, 110, 158, 167–69
media, 16, 32, 91, 96

Meghna, *14*, 15, 18, 93, 98, 182
Mekong River Commission, xxv, xxviin10, 51, 57, 180–81
Middle East, 4, 5, 9, 76, 143
Ministry of Environment and Forests, 72, 128
Ministry of Water Resources, 128
modeling, 29–30

National Commission for Integrated Water Resource Development, 137
National Commission for Reviewing the Working of the Constitution, 127
National Rehabilitation and Resettlement Policy for Displaced Persons, 43
National Water Development Authority, 90
national water policy, xxii, xxvi, 24, 29–34, 45, 79, 89, 134, 137, 149, 154. *See also* policy development
National Water Resource Council, 24
navigational uses, 18, 49, 51–54, 61
naxalism, 44–45
NCRWC, 127, 136
Nepal, 45, 53, 55–56, 58–59, 60, 65, 80, *91*, *92*, *93*, *95*, *97*, 111, 117, 123, 125n17, 141, 144–45, 151–52, 155–56, 175, 181–84
Netherlands, 51, 109, 155
New York Resolution on Flood Control, 53
NGOs, 8, 10, 43, 54, 67–68, 95, 123–24, 136, 140n8, 151–54, 160, 162
Nile Basin Treaty, 60
North-South: A Programme for Survival, 7

OECD, 30, 50–51, 117
opportunity costs, 35, 77, 123
Orissa, 31, 44, 97–98
Our Common Future, 7

Index ~ 229

Pakistan, xi, xxv, 5, 10, *11*, 13, *14*, 15–16, *17*, 18–19, *33*, 41, 53, 56, 59, 60, 65, *80*, *81*, *82*, 83–84, 92, 122, 142, 144–45, 151–52, 156–57, 165, 175, 178, 181, 184–87
Parliament, 43, 128–30, 134, 137, 139, 151–52, 161
Peninsular rivers, *xxxiv*, xxvi, 13, 15, 89–90, 92, 107, 115
policy development: domestic, xxi, xxiii, 6, 26, 31, 91, 136, 149, 152–53, 159–60; international, 94, 105, 107, 109, 127, 141–42, 144, 146, 152–55, 161, 171, 175, 179. *See also* stakeholder engagement
pollution, 8, *27*, 28, 35, 51, 53, 55, 61, 66–67, 69, 76, 99, 106, 137, 155. *See also* contamination
population, xi, 12, *17*, 79; displacement in other countries, 38–39, 47n16, 47n21, 163n1; domestic displacement, xiii, 37–45, 47n17, 75, 99, 151–52; growth, 3, 13, 16, *17*, 52, 63–65, 79, 80, 129, 131, 155, 184; statistics, xi, xxiv, xxvin3, xxxiii, 5, 12, 16, *17*, 20n6, 21n19, 25, 40, 56, 59, 78, 87n20, 151
poverty, xviii, 7, 13, 26, 38, 44–45, 59, 110, 121, 132, 146, 151, 176–77, 182
project: assessment, xxii, xxiii, 41, 71–72, 90–91, 100, 109, 115, 123, 128, 131, 167, 180; culture, 110; funding, 9, 41–42, 74, 109, 120, 122–23
property rights, 26, 29, 39, 46n6, 61
Public Accounts Committee, 71
public domain, 43, 96
public interest litigation, 137–38

rainfall, xxi, 4–5, 15, 17, 28, 67, 79, 89, 95, 100n1, 109, 132, 150, 184
rainwater harvesting, xiv, xxii, 17, 26, 67–70, 85, 127, 142, 144

recycling, 67, 78–79, 85
regional considerations, xxiii, xxvi, 28, 51, 57, 61,141, 145–46, 150–51, 154, 182, 184, 187
Rhine, 28, 50–51, 55, 61
rice growing, 4, 35, 37, 84, 85, 153
Rio Conference, 7, 152–53
River Boards Act, 136
River Linking Project, xii, 28, 89–100, 107–8, 115–24, 138, 145, 162, 171–72, 178
rivers, 13–15, 89; conventions and regulations, 49–62; economic importance of, 52; importance as transport routes, 52; restoration of, 93
Rourkella steel plant, 38

Salal hydroelectric project, 59
salinity, 30, 80, 83, 94, 98, 132–33, 158
Sandoz factory, 51, 61, 155
Sardar Sarovar Dam project, 9, 37, 40, 70, 72, 75, 152
Sarkaria Commission, 136, 139
Scheduled Tribes and other Traditional Forest Dwellers Bill, 43
sedimentation, 72–73, 97–98, 101n13, 108
siltation, 18, 68, *73*, 74, 101n14, 102n17, 168. *See also* sedimentation
sloping land, *73*. *See also* erosion
socioeconomic conditions, 17–18, 31, 38, 136, 139, 146, 159, 168, 175–76
soft law instruments, 44
soil degradation, 73–74, 83, 89, 132, 150, 158
South America, 73
South Asian Association of Regional Cooperation, 51, 118
Southern African Development Community Region Protocol, 51
sovereignty, 50, 52–53, 137, 158
Snowy Mountain project, 100

Spain, 61, 95, 152
Sri Lanka, xi, 45, 124, 167, 169,
stakeholder engagement, xxii, 6, 25–26, 31, 34–35, 90–91, 97, 107, 122, 133, 136, 149, 151–53, 158–62. *See also* transparency
state government, 26, 30, 42, 86, 90, 119–20, 125n9, 127–28, 130, 138
Stockholm Declaration, 50
subsidies, 154
Supreme Court, 39, 119, 128, 131, 138, 145, 152, 187
Sundarban forest, 99
sustainable use of water, xiv, xviii, xxvi, 6, 25, 32, 79–80, 149, 153, 160–61, 171

Teesta River, 57
Tehri Dam Project, 37
temperature, 4, 35, 66, 94, 105, 109–11, 150, 162
Tennessee Valley Authority, 8–9, 20n7
Tipaimukh dam, 57
Tirthas, 23
transparency, 35, 96, 124, 163n7. *See also* stakeholder engagement
Tribals, xiii–xiv, 38, 40, 43, 156, 166, 177, 180, 185
Tulbul project, xxv, 59
Turkey, 53–54

Unemployment, 121, 178
United Nations, xxi, 5–6, 76, 93; 1997 resolution, 52–54; Committee on Natural Resources, 7; conventions, 49, 51–55, 60; Environment Programme, 105; Framework Convention on Climate Change, 106; guidelines, protocols and reports, xxvi, 6, 7, 44, 61, 94, 127, 141–43, 150–51, 154, 171, 181
United States, xxiv, 5, 8, 26, 54, 111, 123, 132, 152, 155–56, 185
Universal Declaration of Human Rights, 44
use efficiency rate, 84, 86

Vested interests, 38–39, 69, 91, 96, 111, 133, 139, 149, 188n9

water markets, xxviin6, 76
water policy, xxi, 23–45, 79, 85, 89, 91; and displaced people, 34–35, 37–44, 173. *See also* national water policy
water pricing, 17, 19n4, 30, 46n5, 49–50, 71, 78, 85, 153–54
water sharing, xxiv–xxvi, 18–19, 24, 55–58, 61, 130, 136, 175–78, 181–82, 184, 187; agreements, xxv, 16, 24, 49, 55–60, 76, 131, 139n2, 144–46, 150–51; disputes, xxviin5, 21n18, 129, 130–31, 135, 144
Wild Life Protection Act (1972), xii
women, xxii, 6, 25, 31, 69
World Bank, xii, xxi, xxv, 5, 7, 20n9, 56, 64, 74, 117, 122, 142, 152, 156, 167, 169, 171, 183; policy shortcomings, 7–9, 161; project evaluation failures, 41–42, 71
World Conservation Union, 111
World Health Organization, 69
World Meteorological Organization, 105
World Trade Organization, 36, 123
World Water Development Report, xi, 171
World Wide Fund for Nature, 109

About the Author

Binayak Ray is a visiting fellow in the department of political and social change at Australian National University.